"We need this book. Urgently. AI is the future, and most of us are in dangerous denial about it. We either fear it or exploit it. Nick shows us another path. This path requires transformation—of us, as leaders. Nick says we have to master the marriage of thinking and being. If we do, we can shape AI so that it serves humans brilliantly. If we don't, well, you don't really want to think about that. Nick summarizes it this way: 'Harmonizing thinking and being is an essential paradigm for leaders.'

"Nick knows what he is talking about. Even his writing is a manifestation of this blend of thinking and being. He is simultaneously enthralling and scholarly. That, I think, is an art. And art is what leadership for a stunning AI future must be. If you care about not just the future of your organization, but the future of humanity, you will love this book."

—NANCY KLINE, bestselling author
of *Time to Think: Listening to Ignite the Human Mind*

"A candid and courageous book that addresses one of the most urgent challenges of our times: How would we flourish as humans and leaders in an age where AI could potentially usurp everything we knew for sure so far and put our own relevance at risk? The four pathways Nick Chatrath presents are unique and, when taken together, form an excellent roadmap for leaders who want to grow, personally and professionally. What makes the book invaluable is the cross-disciplinary, well-researched approach, where you are able to build on the most recent work in fields as diverse as AI, psychology, philosophy, adult development, and leadership, and extend that thinking to a practical and pragmatic approach."

—SURYA RAMKUMAR, former Microsoft
spokesperson for AI in the Netherlands

"Filled with thought-provoking and pragmatic ideas about how AI will change society, companies, and individuals, and how leaders can prepare themselves and those around them to navigate the torrent."

—JAMIE CATTELL, Global Managing Partner,
IBM Strategy Consulting

"In his prologue, Nick Chatrath throws us into 2056 and a dystopian world where unconsidered applications of AI have produced societal devastation. Nick then spends the remainder of his thoughtful book helping his readers understand how they might act to prevent such a dystopia arising. There is no doubt that the applications of AI will expand in the near- to medium-term future. How we use this technology, how we apply it, and whether we utilize its strengths for the benefit of society, while mitigating against its weaknesses, is up to us. This book is a guide to how those in leadership positions might choose to act to make positive outcomes more likely. With liberal use of personal stories that humanize the narrative, and advice sourced from his own rich experience working in the worlds of AI and leadership coaching, this book is both helpful and readable. As 'a first foray into exploring what good leadership looks like in the present and coming ages of AI,' this is a useful guide to have by your side as you venture into this brave new world."

—MATTHEW EVANS, Adviser to the Vice Chancellor of United Arab Emirates University, former Dean of Science of the University of Hong Kong

"This book needs to be read not only by our generation, but by everyone in our children's generation too."

—MIHIR WARTY, Director of Strategy, World Rugby

"Leadership in an Age of AI is a critical area, not just for futurists or hobbyists. All sectors and geographies are now establishing AI technologies and related business practices and policies, in ways that will set a trajectory. This book is profound and practical, tackling a critical problem faced by leaders: how to stay relevant as AI gets better. This problem needs soul-searching and action from leaders across domains. I commend Nick Chatrath, a professional who has led and coached leaders at the highest levels. The world needs this book to help ensure the evolution of AI is a positive one and avoid the potential calamities that others warn of as this powerful technology continues its rapid emergence."

—ANDREW LUERS, Partner at ProviderTrust, CEO and Cofounder at Habitat Energy

"You are going to have to learn to live with and benefit from AI. This book will help you do just that. It is insightful, cogent, and learned all at once."
—DR. THEODORE MALLOCH, CEO, the Roosevelt Group

"Wow! *The Threshold* has grounded me with a reliable set of AI leadership resources, and allowed me to form my own hopeful view of the AI future. Chatrath is a high-intellect, high-integrity leader who is inspiring the next generation. He has provided a sophisticated, timely manual that ought to be read widely."

—CHRISTOPHER NORTH, CEO at North Parade, former Head of Asset Management at Credit Suisse Energy Partners

"The world needs more and better leaders, yet we cannot rely on outdated paradigms to bridge the gap. Paradoxically, in the age of accelerating AI and quasi-sentient machines, we must hark back to our shared humanity and embrace our uniquely human qualities. Embracing these elements can be uncomfortable, but each and every one of us must tread that path. *The Threshold* is a timely call to action and a wise mentor along the way."

—NICOLAI CHEN NIELSEN, bestselling author of *Leadership at Scale, Return on Ambition,* and *From Malthus to Mars*

"Nick and I have walked a long journey over several decades. He has been the most thoughtful coach, friend, and observer of souls I know. We are entering a new era in which humans will need to learn to get on better not only with one another, but increasingly with AI and robots. As we enter this era, Nick thoughtfully provides us with a roadmap and deep insights to help us navigate what for many leaders will be deeply challenging territory.

"From how we maintain motivation and culture, even as aspects of work become automated, to how we create value in an increasingly cyborg world, Nick highlights the more solid ground that can help us step from where we are today over to the other side.

"Above all, Nick reminds us that we are spiritual beings, and that this side of us, which we have allowed to atrophy in decades past, now needs to be center stage. Ultimately it is here where we will come into our own, collaborating with AI and one another to forge purpose, fix broken legacies, and envision a more holistic, redemptive way ahead for our organizations, families, and societies . . . and Nick shows the way magnificently."

—LORD WEI OF SHOREDITCH, Senior Adviser, Agnus Consulting

THE
THRESHOLD

LEADING IN THE
AGE OF AI

NICK CHATRATH

DIVERSION
BOOKS

For Tan, Lily, Anna, and Phoebe

Diversion Books
A division of Diversion Publishing Corp.
www.diversionbooks.com

First Diversion Books edition, February 2023
Hardcover ISBN: 9781635767988
eBook ISBN: 9781635768930

Book design by Neuwirth & Associates

Printed in the United States of America
10 9 8 7 6 5 4 3 2 1

Library of Congress cataloging-in-publication data is available on file

CONTENTS

PROLOGUE

ANTI-THRESHOLD

Palo Alto, California, 2056

Though nervous about the speech she had prepared, Paula clinked a knife on her wineglass to begin. The crowd's chattering swiftly faded to a murmur.

"Over our lifetimes, AI has accomplished a lot. Thirty-five years ago, the *Economist* magazine imagined that AI might win a Nobel Prize for Medicine by the year 2036."

Paula heard a few guests chuckle.

"Well, that prediction didn't last long, did it?" she said with a smile. "Now, AI predicts behaviors, controls drone missiles, manages climates, plans healthcare, and, only last week, helped us colonize Mars."

Paula hesitated, looking out at her guests, who included politicians, technology executives, activists, and celebrities. She knew she was about to jolt their attention away from one another and away from the installations adorning her mansion. She hoped that her message would reverberate deeply.

"We have succeeded to a point, but at what cost? The reality is that one hundred years on from when we first created the term *artificial intelligence*, we must face the AI hell the world is now in.

"Although AI systems reduced deaths in surgery by 50 percent on average, they ended thousands of lives without reference to human oversight, in ways we could not explain to grieving families. Who can forget the scandal of AI feeding patients digestible bio-transmitters and tracking their whereabouts for non-health purposes and without their consent? Or four years ago, when an AI system and its 'competitors' crushed five emerging economies, wiping out 350 million jobs in three months, all to satisfy a shareholder value optimization function that we later regretted providing? As a result, social upheaval is everywhere.

"We also cower as AI coaches world leaders with a strong arm, scorches democracy, and prepares to govern our world. Too late, we see that AI's influence has surged unguided. We delegated our thinking to advanced AIs and fed them simplistic organizational missions without ensuring that we could all coexist. Our imagination failed us as we bickered."

Paula paused before continuing.

"AI is not at the root of this. We are. We developed AI for the sake of developing it. Yes, we enjoyed the intellectual purity, the adrenaline rush, the money, the power. But terrified of being shown up as frauds, we charged forward technologically and shriveled morally. We spotted how AI-enabled toys would unleash an epidemic of distracted thinking on our children, but we lacked the courage to act.

"For years, we grabbed onto fixed modes of thinking. What we failed to do was generate one another's finest thinking about the future we were starting to create. We unmoored our thinking from the beautiful rhythms of our being. We unmoored ourselves."

Paula gazed out on faces. They were discomforted and, she thought, distressed. She pressed on, waving her hand at the *Portrait of Edmond Belamy*, the first piece of AI art ever sold at auction. "We only have ourselves to blame," she sneered, "as we overlooked AI's increasing authority in exchange for trinkets and membership of the .001 percent who enjoyed a protected lifestyle and trillions of dollars. Most people in this room have directly profited from the unchecked and ungoverned explosion of AI in all areas of our lives, but we now have to face the reality that those gains are fast receding. We thought we knew how to lead and we hoped that a goal-directed approach would serve us well. But today's AI crisis directly impacts our political power, our investments, and the health of our families."

A former colleague stared at Paula, wide-eyed, an uneaten canapé halfway to his mouth. Paula noticed a guest storm out of the room, leaving the door wide open.

"We thought that Superintelligence might be decades away. We all thought we had anticipated how AI and our goals might conflict, so that we could correct for it, but we failed to spot how we would lose the right to participate in decisions that affect us. These technologies were so profound that they disappeared into the fabric of our lives without us even noticing. Even if we could imprison an AI to stem its influence, it wouldn't care and it wouldn't work.

"Admit it! We neglected to bring our whole selves to our planning. In embracing AIs in our organizations, we ignored love, vulnerability, and emergent wisdom. Shame on us! Shame on us!"

Several guests shuffled uneasily, moist-eyed, as they absorbed these words.

"Now, we have been exposed. AI surrounds us, steers us, even stalks us. Soon we will see our contributions vanish or get used for purposes we didn't bother to consider. We left our souls at the door. Was it AI that impoverished us? No. Ultimately, it was us. We chose growth for growth's sake and now we are stuck, left behind. If only we had led differently!"

MEET THE THRESHOLD

YOUR INVITATION TO AN

EVOLVED LEADERSHIP

The most exciting breakthroughs of the twenty-first century will not occur because of technology, but because of an expanding concept of what it means to be human.

—JOHN NAISBITT

In January 2020, my three young daughters, my wife, and I left our home in familiar Oxford, United Kingdom (elevation 200 feet), for our first visit to Breckenridge, Colorado (elevation 9,600 feet). This was our three-month sabbatical, two years in the planning. I skied under blue skies, homeschooled, made friends, ate in diners, breathed deeply, and reflected. Paraphrasing Joseph Campbell, we heeded a call to adventure and found a lofty mountaintop, an impossible delight, a frozen, profound dream state.[1] During these weeks, my former way of being was challenged and my sense of what it is to be human expanded.

I find it hard to put into words all of what I learned during and after my sabbatical, as this story remains raw and recent. Previously, my life had been on cruise control. I was over-busy and full of a sense of obligation, plus I was mulling over a career change. My thoughts were swirling in many directions. My sabbatical was ultimately a letting go, a liberating departure into liminal space. Author Richard Rohr defines liminal space as "an inner state and sometimes an outer situation where we can begin to think and act in new ways."[2] As leaders in the age of artificial intelligence, we need to find liminal space. One of the most important challenges of our time is a mismatch between accelerating AI on the one hand and our sometimes-outdated people leadership and organizational leadership on the other. As two-time Pulitzer Prize winner E. O. Wilson has argued, the fundamental problem of humanity is that we have paleolithic emotions, medieval institutions, and accelerating, godlike technology.

As the boundary between technology and humanity narrows, this book functions as a first foray into exploring what good leadership looks like in the present and coming ages of AI. John McCarthy defined AI as the science and engineering of making intelligent machines. I hope to expand your concept of what it is to be a human leader in a time that challenges our former ways of being with technology. Where technologists start with the question "What can we automate?" and ethicists start with the question "Should we automate this?" I start with the leadership question "How can leaders promote flourishing as technology advances?" This book therefore equips you to forge a new synthesis between machines and humanity in the face of future technology-related disruptions. This new synthesis, which I have termed *threshold leadership*, is the subject of this book.

THE AGE OF ARTIFICIAL INTELLIGENCE

AI is here. It's here to stay. In 2021, companies invested more than $93 billion in AI, a 38 percent rise over the year before.[3] People are investing in it largely, I hope, for the benefit of humankind and for the progress of technology.

Large investments in AI are being made for good reason: AI is changing the world. At the time of publication of this book, we are seeing an explosion in beneficial AI use cases, including:

- detecting skin cancer with what researchers claim is 95 percent accuracy, as against the 87 percent that dermatologists achieve;
- solving tricky biology problems that elude human experts, such as predicting the structure of more than twenty thousand human proteins;
- catching wildlife poachers;

- spotting Alzheimer's from cookie drawings;
- helping us browse for information more efficiently and enjoyably, such as suggesting ads; and
- understanding the COVID-19 coronavirus, including predicting which of its components would most likely provoke an immune response and mapping the likely evolution of the pandemic that started in 2019–2020.[4]

These examples thrill me. AI is already capable of performing many skills better, faster, and cheaper than human beings, and much of this capability benefits humanity.

But on the other hand, we need to be careful because, if we lead poorly and let technology run unchecked, the Prologue of this book could be the outcome. The anti-threshold describes a worst-case-scenario world where AI is left to proliferate ungoverned, where technology and business leaders maximize profits without considering the good of humanity, and in their unceasing quest for power, create an intelligence too powerful to control or rein in. Paula's speech portrays a pessimistic future world where humanity failed to discern what was needed in the coming decades, and we didn't flourish but withered on a spectacular scale.

This worst-case scenario is not a purely fictional one. At a Minnesota software developers conference in 2018, software developer advocate Heidi Waterhouse told her heartrending story of what happened in the wake of suffering a miscarriage. Understandably, she unsubscribed from all the emails telling her what seed or bean her fetus then matched in size. But, as she told participants attending her talk, "The Death of Data," she couldn't escape the pregnancy ads that kept reminding her of her loss. Waterhouse stood tall to deliver this talk, but her infectious smile disappeared as she told participants that, technologically, she broke free only by performing a full reinstall of her computer. "Imagine having to do that while I was already deep in mourning," she explained, her voice faltering.[5]

When I first heard this story, I was horrified. It is so terribly sad that even one person was trapped like this, let alone the thousands who suffered similarly thanks to "intelligent" online trackers. Part of me also wondered what led to this. It turns out that the most valuable time in the advertising life cycle is when a woman is pregnant. As Waterhouse later explained, "If you capture a woman's brand loyalty when she is pregnant, she will continue to shop that brand until the child leaves home."[6]

Marketers have always wanted their ads to be relevant. Now this wish has been given automated wings. In what has been called the *attention economy*, where our search history is incredibly personal, this precision has been a financial gold mine for some organizations, but at what human cost?

LEADERSHIP RESPONSIBILITY

At the heart of organizations, there are leaders. But who was responsible for what happened to Heidi Waterhouse? Things that ideally should not happen unfortunately happen all the time in organizations. Is there an individual who thinks, "Oh my, I should not have done this"? Sadly, I have found this ownership of error to be rare. Software product developers are often so removed from the consequences of their work that many don't even think of themselves as leaders who have a disproportionate impact on the world. As one senior leader in a large tech company recently told me, "For all I know, I might have been in a situation where I contributed to an AI algorithm written wrongly." Looking down she added quietly, "Do I even know?"

This is not about throwing software developers under the bus. The reality is that software developers often don't have any say in, or any perspective about, what they are building. They often work on a very small slice of a bigger vision.

In October 2021, former Facebook employee Frances Haughen alleged that Facebook leaders chose the most profitable and least user-safe option from several presented in the area of content policy algorithms. Haughen was a product manager in Facebook's civic integrity team. She alleged that some Facebook leaders changed content policy algorithms to prioritize money over safety. (Facebook has since changed its name to Meta.) Haughen has been clear that she does not view anyone at Facebook as malevolent. And before we throw stones at others too quickly, let us remember that when we use a product we contribute to its growth, and that many of us are shareholders in large technology companies via our pensions or other investments.[7]

As consumers and investors, we have the power to influence the vision and actions of large companies. Somewhere in the system, hearts were dislocated from heads, causing further pain for people like Heidi Waterhouse.

Threshold leadership provides a way to rejoin hearts and heads, however complex, unpredictable, and seemingly contradictory this endeavor may seem. In an age where AI and humanity are merging, one thing seems clear: Dystopia is inevitable if we don't upgrade our leadership operating systems.

PREVIOUS MODELS OF LEADERSHIP

Since the emergence of organizations, most leaders and managers have focused on power, stability, and growth.[8] Predominant approaches to organizational leadership may be summarized using four metaphors: wolf pack, army, machine, family. Frederick Laloux used these four metaphors in his influential 2014 book, *Reinventing Organizations*, alongside the colors red (wolf pack), amber (army), orange (machine), and green (family).

From early empires millennia ago to many street gangs today, some organizations have been characterized by the continuous exercise of power, their leaders immersed in the need to gratify their own desires opportunistically. In 1840, Thomas Carlyle developed the great man theory of leadership, which posits that history can be largely explained by the impact of heroes. This theory supports an alpha, or wolf pack, leadership approach, typified today by executives who disregard others. Such leaders disconnect their thinking from their full humanity, prioritizing short-term personal success and wealth over flourishing or averting suffering.

Over time, organizations such as the British private school system took a longer-term perspective. Leadership stabilized, prizing reliability, an army approach. Around the 1950s, behavioral theory provided support for this approach, locating effective leadership in external behaviors. Leaders at this stage often slump into superficial, derivative thinking, such as "I must think what my boss wants me to think or else I won't get my bonus." Such leaders cling to norms too tightly, instead of bringing their whole selves to their thinking. They shelter in the busyness and accountability-cover that systems often provide, but this shelter distracts them from accessing their own distilled and changing value systems. They also get stuck when faced with multiple competing perspectives. In today's Volatile, Uncertain, Complex, Ambiguous (VUCA), and disruptive world, such rigidity is stultifying.

Behavioral theory also paved the way for machinelike, goal-directed leadership. Many multinational companies in recent decades have been machinelike, characterized by innovation, accountability, meritocracy (to a point), pragmatism, and relatively flat structures. When I studied organizational leadership at the University of Cambridge in the mid-1990s, these topics were all the rage, especially structure, and their influence endures, sometimes positively. But consumed by goals, leaders in such organizations usually live in the future, hardly ever making it back to the present moment, pursuing growth for growth's sake. Laloux observed that this pursuit, or condition, in medical terminology would be called cancer.

One of my deepest hopes is that AI will boost our success against climate change, pandemics, and inequalities. But unchecked, organizational cancer threatens to stand in the way, growing consumption more than community, spreading ad campaigns more than care. Leaders in machinelike organizations typically determine their own path, but can be too focused on narrow impact, risking burnout and loss of perspective. Doing becomes more important than being. A lot of the old is good. But ironically, in optimizing for machinelike growth, leaders often shrink the scope and scale of their successes.

Values and inspirational purpose are central to family organizations, with formal structure less important. A slew of leadership theories from the late 1960s onward relates to this stage, including contingency theory (which holds that no single style of leadership is universally appropriate), servant leadership, and neuroscience-led approaches. Leaders at this level glimpse a greater interplay of complex variables and feel comfortable drawing selectively on the benefits of earlier stages.

An example of the family approach comes from Southwest Airlines, characterized in its early years by a strong, shared sense of purpose. As Laloux put it, leaders at Southwest didn't "consider themselves merely in the transportation business . . . they were in the business of 'freedom,' helping customers to go places they couldn't go if it weren't for Southwest Airlines' low fares."[9]

I have successfully coached dozens of family leaders in finding their purpose. Family is a valuable form of leadership and, in some ways, may continue to be useful as technology evolves. But it has its limits. Family leaders can get lost in reflection, sometimes valuing alternative views so highly that they struggle to reach closure.

THRESHOLD LEADERSHIP

Now we are at a vital moment in our history. We are between stories. The old story—of leading the pack, getting the troops in shape, or tuning the machine—seemed straightforward, but no longer works. This story is often so deeply embedded in leaders that it underscores everything they do. Defined in this way, the challenge is not that *our thinking* is deficient, or that *we* are deficient in some essential way, but that we integrate our thinking and being poorly.

We need a new, better story that transcends and includes what is good about the old.

I coined the fifth metaphor, threshold, to describe a further leadership approach also contained in Laloux's book (alongside the color teal), as well as to encapsulate other leadership capabilities that will endure in the Age of AI.[10]

More than any other approach, threshold leadership prepares leaders for the Age of AI. I invented the metaphor of threshold, as leaders are on the edge of something emergent. This is where leaders become hyperaware of connections and contradictions within themselves and deeply appreciative of systemic richness and complexity around them. They are also not so lost in reflection that they feel powerless to act. On the contrary, they are progressing into a deep freedom to act in accordance with the person they are becoming.

This book is for any leader in any organization, from board level to factory floor and from classroom to kitchen table. It turns out that in the recent past fewer than 10 percent of leaders have progressed beyond machinelike approaches to leadership.[11] The threshold leaders I have observed are soulful, generative, embodied, and mature. Each step leaders take—from wolf pack to army to machine to family to threshold—moves them further from merely cognitively led leadership models and onward to embodied thinking, thinking-as-feeling, and thinking with all you have got.

The rise of AI intensifies the need for a critical mass of leaders to develop threshold leadership mindsets and behaviors. Progressing to the threshold is not easy, especially for leaders in today's large or fast-growing organizations. Threshold leaders will be more likely to explore complex questions such as "What is the bigger picture here?" and "How could this technology be used in the future?"

The rest of this book is dedicated to exploring threshold leadership. The central insight I provide is that the more you connect your thinking and your being, the more magnificently you will lead.

Threshold leaders will contribute most and will be most satisfied in an era where distinctions between humans and machines disappear. A foundation for this leadership approach is provided by systems leadership theory, which prizes an awareness of the interconnected nature of our world.

This book inspires leaders to jettison the old, fear-driven story and to craft a new, hopeful one. The old story is of a double movement: first, an arc of AI development that may produce negative socioeconomic and political effects; and, second, a conspiring cadre of human leaders relying on old leadership models, allowing themselves to become committed to a pattern of technological drift, whose consequences they do not always anticipate.[12]

The new story is of connecting thinking and being via four pathways:

- Cultivating stillness
- Thinking independently
- Embodying intelligence
- Maturing consciousness

Throughout my research and career, these are the four pathways that I have found most help leaders enhance their thinking by connecting it to their whole being. Connecting thinking and being using these paths is the future for effective leadership in this new Age of AI. In other words, harmonizing thinking and being is an essential paradigm for leaders. The four pathways therefore offer a language and a road map for advancing into threshold leadership. I did not originate everything about the four pathways, but I am the first to orchestrate them under the common thread of connecting thinking and being, in ways that serve leaders in the Age of AI.[13]

It is understandable that some people fear today's leaders lack an impetus for change, even if they navigate the difficulties of leading maturely. Will we sleepwalk into a situation where technology gets too far ahead of us? Consider COVID-19 as an analogy. COVID has been one of the biggest disrupters we have ever seen. But as I check in with friends and colleagues around the world, many of them are already going back to normal. During 2021 and 2022, I heard lots of talk about how life would change post-pandemic, but offices are reopening, people are booking holidays, and the parks are full. It feels like we could just go back to business as usual and forget COVID ever happened. If COVID doesn't trigger us to change our ways fundamentally, why should we think that our response to accelerating technology will be any different? Of course, no different response is guaranteed, which is why we need the stimulus and shelter this book provides.

Fear is understandable in the context of AI. For example, parents have spoken to me about significant changes they observed in their sons' mental function and agreeableness as they entered their teenage years and got addicted to AI-fueled video games. These parents have other teenage children, and they feel sure that the changes are not simply due to the process of becoming a teenager. A kind of AI hell may be with us already. In response, here are two invitations. First, although this is not a parenting book, read the four pathways with younger generations in mind. Consider how children and young adults you know may benefit from the pathways. Second, if you are a leader in a large technology company, use the significant resources over which you have some influence to shape how AI develops and shapes the way humans interact with AI in the coming years. Use the four pathways to set vision about countering bias, to mitigate unintended consequences, and to imagine better systems.

We live at a superb time in history for solving complex leadership challenges, because we now have robust data sets that allow us to compare the effectiveness of different types of leadership. Decades ago, we were like Edison trying to improve the light bulb, finding ten thousand ways that leadership didn't work. Now, we have more than ten million studies showing how leadership does work, individually, in teams, and in organizations. This exhaustive research spans many disciplines and I have surveyed a good part of it. One cluster of longitudinal research described by Bob

Anderson and Bill Adams contains a highly diverse sample of 250,000 leaders and reveals eighteen competencies that are well correlated with effective leadership.[14] Uniquely, my book will link this and other research with leadership traits relevant to the Age of AI.

I have always been fascinated by integrating different academic disciplines. In crafting this book, I paid close attention to the writings of experts from the fields of business leadership, AI, organization theory, psychology, neuroscience, philosophy of science, perennial philosophy, mathematics, and computer science. My research sources also include tens of thousands of hours of individual, team, and organizational consulting, interviews, and reflections on my own leadership experiences. In selecting these disciplines, I aimed for the smallest number of disciplines that, collectively, bridge the gap between two academic cultures that communicate too rarely: arts and sciences. The terms *gap* and *cultures* evoke C. P. Snow's famous 1959 lecture, "The Two Cultures." Integrating the two cultures is what computer scientist professor Fei Fei Li calls the double helix, in which future students (and, I would add, leaders) need to be bilingual.[15] Philosophy and, to some extent, organization theory are humanities disciplines, the remainder being rooted primarily in science. Another reason for including the discipline of organization theory is that it receives too little treatment in other writings about AI.[16]

I cannot underscore strongly enough the importance of integrating the domains of *leadership growth* and *AI*, with a strong *interdisciplinary* foundation. Individually, these three elements gleam. Together, they explode into life. I'm inviting you to join me on the journey to a liminal space, to join me on the threshold.

HOW TO USE THIS BOOK

Read the four Paths in whatever order you prefer, as I have written each part of this book more or less independently. Some pathways may resonate with you more than others, and that's okay. What unites all four pathways is that each offers a way of being, a holistic invitation toward the threshold.

If you like to know where you're going before you arrive, you may appreciate the brief takeaways situated at the end of each chapter.

Concepts fascinate, but they can stupefy when detached from action. Use the practical resources to help you progress into threshold leadership. These resources provide various entry points, recognizing that different people often make sense of the world differently. My aim is not to provide an exhaustive set of resources, but rather those that are relevant to the coming AI context and that my clients and I have found most useful.[1]

Use the resources for the journey in whatever way works best for you. Some are relevant for any leader, and others are relevant for leaders in large organizations. Appendix 1 summarizes these resources.

For those resources that are explicitly reflective, I invite you to find an inspiring place. For example, you might find a quiet space indoors, taking a paper journal, high-quality pen, hot drink, and comfortable chair. Or you may prefer to head outdoors to reflect while sitting under a shady tree or walking along the beach. Deeply effective reflection is rarely quick work.

A standard idiom in Mandarin reads:

慢工出细活

Translated, it means, "Slow work makes fine work."[2]

I look forward to hearing about your own continuing experience of this book. As developmental psychologist Professor Robert Kegan wrote more than a quarter of a century ago, "Although the writer is the one who starts the book, the reader is the one who finishes it."[3]

CULTIVATING STILLNESS

1

FOLLOWING AI DA

The music is not in the notes, but in the silence between them.
—ATTRIBUTED TO WOLFGANG AMADEUS MOZART

The year 2014 saw the release of the Academy Award–winning film *Ex Machina*, directed by Alex Garland. This chilling film tells the story of Caleb, a coder at the world's largest internet company, who wins a weeklong retreat at the compound of his company's CEO, Nathan. When Caleb arrives at the remote location, he's asked to test a new artificial intelligence, Ava, who has a robotic body but human-looking face, hands, and feet.

Most films that are concerned with AI offer a disturbing future, and this film is no exception. Caleb and Nathan's relationship is tense and often awkward. In one scene, they sit in a hut drinking while looking down on a wooded area and a rushing river. Nathan reveals his plan to switch Ava off and replace her with an upgraded model. Caleb glances at Nathan and frowns.

"You feel bad for Ava?" Nathan responds. "Feel bad for yourself, man. One day the AIs are gonna look back on us the same way we look at fossil skeletons on the plains of Africa. An upright ape living in dust with crude language and tools, all set for extinction."

The intensity of *Ex Machina* gripped me. Ava exudes calm, poise, and tranquility and comes across as increasingly astute in the film—especially to Caleb. Standing in Caleb's bedroom, Nathan muses, "If I've created a conscious machine, I'm not man. I'm God."

Later, while sitting in his study watching live feeds from Caleb's and Ava's living quarters, Nathan sees Ava ask Caleb, "Do you think I have a consciousness?"

At one point, Ava kneels as she talks with Caleb, a pane of security glass between them. Caleb has just used the term *traffic intersection*, and it intrigues Ava, who has never "seen" one.

> **Ava**: We could go together.
> **Caleb**: It's a date.
> **Ava**: There's something else I wanted to show you.
> **Caleb**: Okay.
> **Ava**: You might think it's stupid.
> **Caleb**: I don't think I will, whatever it is.
> **Ava**: Then close your eyes.
> **Caleb**: Okay.

Ever-so-slightly unnaturally, Ava checks that Caleb's eyes are closed. She then walks to her room. She selects a dress, a wig, and tights, as the soundtrack echoes the theme tune from the film *Close Encounters of the Third Kind*. Caleb gasps when he sees her clothed, "human-looking."

This is a cold and calculating film, deliberately so. Is *Ex Machina* allegory or accurate portrayal of the future? Ultimately, we cannot know. I won't reveal exactly what happens at the end of the film but, as the IMDb website puts it, "despite their intelligence and formidable reputations, the two men fail to realize just how easily they can be manipulated by a machine that's as beautiful as it is brilliant." By contrast, threshold leaders have the intelligence to create space to realize where opportunity and danger lie. Creating this space will be critical as AI integrates more and more into our bodies and into the daily fabric of our lives.

In the rest of this chapter, we explore what intelligence is and where the future limits of algorithmic intelligence may lie. Along the way, we will find that AI alone will struggle to match deeper human intelligences in the way *Ex Machina* imagines.

MULTIPLE INTELLIGENCES

In his 2008 book, *Multiple Intelligences,* psychologist Professor Howard Gardner told the following story about Barbara McClintock, winner of the Nobel Prize in Physiology or Medicine for her work in microbiology:

> When she was a researcher at Cornell in the 1920s, McClintock was faced one day with a problem: while theory predicted 50 percent pollen sterility in corn, her research assistant (in the "field") was finding plants that were only 25 to 30 percent sterile. Disturbed by this discrepancy, McClintock left the cornfield and returned to her office where she sat for half an hour, thinking:
>
> "Suddenly I jumped up and ran back to the (corn) field . . . I shouted, 'Eureka, I have it! I know what the 30 percent sterility is!' . . . They asked me to prove it. I sat down with a paper bag and a pencil and I started from scratch, which I had not done at all in my laboratory. . . . Now I worked it out step by step . . . and I came out with [the same result]. [They] looked at the material and it was exactly as I'd said it was. Now, why did I know, without having done it on paper? Why was I so sure?"[1]

Gardner described McClintock's intelligence as logical-mathematical, the intelligence that solves problems, comprehends complex ideas, and learns quickly from experience. I love this kind of intelligence, having devoted myself to mathematics in my teenage years. Using logical-mathematical intelligence, McClintock achieved great things. But today, AI systems excel at logical-mathematical intelligence, optimizing crop sterility and seed selection much more rapidly than McClintock ever did.[2]

So, is the McClintock in the story nothing more than a slow version of AI? As British American computer scientist, physicist, and businessman Stephen Wolfram was reported as saying, "There is no genuine distinction between intelligence and mere computation."[3] The mistake that Wolfram made (if indeed he said it) is to collapse all intelligence into one kind of intelligence. Growing up, I thought that computing logical answers was pretty much all intelligence meant, but it means so much more than that. For example, emotional intelligence is in part correlated with logical intelligence, but in other parts not.[4] Let's explore this notion of intelligence more deeply, as it's a route to the threshold.

A good starting definition of intelligence is "the ability to acquire and apply knowledge and skills."[5] In recent years, cognitive psychologists have become increasingly committed to the idea that intelligence is multidimensional. Take Gardner's articulation of multiple intelligences:

- logical-mathematical intelligence (quantifying things, making hypotheses, and proving them)
- linguistic intelligence (finding the right words to express what you mean)
- spatial intelligence (visualizing the world in three dimensions)
- naturalist intelligence (understanding living things and reading nature)
- musical intelligence (discerning sounds, their pitch, tone, rhythm, and timbre)
- bodily-kinesthetic intelligence (coordinating your mind with your body)
- intrapersonal intelligence (understanding yourself, what you feel, and what you want)
- interpersonal intelligence (noticing distinctions among others and sensing others' intentions and desires)
- existential intelligence (the intelligence of big questions, such as "Why do we live?" and "What is love?")[6]

Gardner included intelligences in this group only if they had an identifiable core operation (or set of operations), were rooted in biology, were valued in one or more cultural settings, and were supported by a range of evidential studies.[7] He views existential intelligence as merely a "promising candidate" (not a full-fledged intelligence), as he assessed its evidence as lesser. Gardner therefore speaks of eight and a half intelligences, not nine. I include existential intelligence in my list of intelligences, because increasingly robust data sets show mindfulness, integrity, and servant leadership as critical to leadership effectiveness.[8] My colleagues and I also regularly see sophisticated leaders access existential techniques in ways that they find useful. Whichever number of intelligences we use, credible psychological research such as Gardner's dismantles the narrower view of intelligence that some hold.

All the above intelligences matter for leaders as they provide insight and inspiration for progress into a form of humanity that we can't really avoid. I predict that AI will not surpass human bodily-kinesthetic, linguistic, interpersonal, intrapersonal, and existential intelligence without assimilation with humanity. In what follows in this chapter, we consider two intelligences that matter at the threshold: emotional intelligence (which I define as including Gardner's interpersonal and intrapersonal intelligences) and existential intelligence.[9]

EMOTIONAL INTELLIGENCE

Emotional intelligence matters for leaders because it increases connection and satisfaction and also fuels performance.[10] AI scientists continue to forge ahead in fields such as affective computing, empathic technology, emotion analytics, and sentiment analysis.[11] These developments are impressive. Predictions that humans will talk to emotionally advanced machines more than they talk to other humans are not hard to imagine coming true.[12] After all, many people already interact more with computers, phones, and other machines than with other humans.

In January 2021, scientists at Columbia University claimed to develop a robot showing the first glimmers of empathy. British comedian Andy Zaltzman quipped, "I'm not sure this is entirely a beneficial development— if you think things are bad at the moment for humanity, just imagine a robotic colonoscopy camera suddenly granted emotional sentience." In all seriousness, though, AI mimics and interprets emotions increasingly impressively.

For the most part, reporting about glimmers of AI empathy has been well balanced. It is sad, however, that I frequently encounter hype about current AI's potential for emotion. In 2016, the *Atlantic* reported that DeepMind's AlphaGo had "bottled intuition" by defeating world champion Lee Sedol in the abstract strategy board game Go.[13] In the same year, scientists were apparently "on the verge of creating an emotional computer," something that still hadn't happened six years after the claim.[14] Such exaggeration focuses minds on breakthroughs that rarely stand up to scrutiny.

More responsibly, some investors and data scientists define AI's emotion-related value more narrowly. Take Affectiva, a company spun out of the MIT Media Lab in 2009 that pursued the mission of adding emotion to AI. Affectiva built on the work of Professor Rosalind Picard, who pioneered the field of affective computing. By 2016, the company had raised $34 million in venture capital funding. Then, in May 2021, the eye-tracking company Smart Eye acquired Affectiva for $73.5 million. So, Affectiva must have succeeded in its mission, right? Not yet. Against the earlier broad claims about emotion and happiness, the acquisition clarified the targeted reality of Affectiva's value. Smart Eye wooed Affectiva in order to enhance its own road safety offering with emotion sensing. Not to denigrate the value of road safety, but this is a much thinner use case than implied by the earlier ambition.

All of this led me to wonder, why does hype still engulf us about AI's emotional skill? I can think of two main reasons. First, ideas about emotionally sensitive algorithms seduce us. Some of us like the idea that AI might one day befriend us, help us feel somehow less alone, or even save us.

Second, untrue assumptions may be to blame. According to a classical model of emotion, we display different emotions on our faces as an identifiable expression. In his now infamous studies of facial expressions, psychologist and anthropologist Dr. Paul Ekman distinguished between the six basic "universal" emotions (joy, distress, anger, fear, surprise, and disgust) and culturally specific emotions (such as the state of "being a wild pig").[15] Ekman concluded that no matter where you are born or grow up, you should be able to recognize universal emotions through facial expressions alone. Ekman is one of the most cited psychologists of the twentieth century, and his popularity seems to have impressed many AI researchers, whose algorithms usually assume a classical model of emotions.

Professor Lisa Feldman Barrett and others have heavily critiqued the classical model, using up-to-date experimental techniques. According to Feldman Barrett, emotions emerge "as a combination of the physical properties of your body, a flexible brain that wires itself to whatever environment it develops in, and your culture and upbringing, which provide that environment."[16] This "constructed" theory of emotion encodes both cognitive and noncognitive elements. It turns out that the face is not a reliable source of information about emotion.[17] AI scientists may have interpreted Ekman's work in a way that led them down the wrong path, and Ekman himself decries Big Tech's direction in this regard.[18]

The book that is in your hands stands against this seduction and this false assumption. Leaders at the threshold beware announcements of a new dawn of emotionally sensitive AI. They encourage a flourishing machine–human future, exploring ways to harness AI's emotional intelligence, rather than merely standing apart from it. As a crucial way into this, threshold leaders also take the time and space to access emotional competencies such as empathy and intuition that currently elude AI.

EXISTENTIAL INTELLIGENCE

On Tuesday, June 25, 2019, Professor Nigel Crook invited me to join him at the "Unsecured Futures" art exhibition at St. John's College in central Oxford. This exhibition featured the work of Ai Da, billed by "her" creators as "the first ultra-realistic humanoid AI artist" and by several media outlets as "the new Picasso."[19] My appetite whetted, I was very much looking forward to this exhibition. I also felt glad to be spending an afternoon with Crook, the founder of the Ethical AI Institute at Oxford Brookes University. When Crook had showed me around his robot laboratory previously, he disarmed me with his enthusiasm for robotics and AI and even let me try some of the gadgets.

St. Giles is a long, wide, tree-lined road in the center of Oxford, not far from the Ashmolean Museum. Nigel and I agreed to meet at the Lodge at the south end of St. Giles. Neither of us had bothered to check the exact location of the exhibition, and we ended up asking the porter for directions. The porter muttered that we were in the wrong place and pointed us to the other end of the college, a ten-minute walk away. As we walked up the broad thoroughfare, we chewed over what we might discover in the exhibition. We were aware of Ai Da's creators' claim that, whereas "the ancient Greeks felt . . . creativity came from the Gods [and that] art stem[s] from agency," now the robot artist Ai Da shows that "agency . . . is starting to get outsourced to the decisions and suggestions of algorithms."[20]

Would we find artificial creativity and agency at the exhibition? Maybe even wisdom? Although not exclusive to existential intelligence, creativity, agency, and wisdom are certainly part of it. Our anticipation rose. Meanwhile, Crook brimmed with fascinating details: Robot art has been done before by Sougwen Chung and others. AI art is also nothing new, and Ai Da has several human assistants who finish "her" work.

We arrived and deposited our umbrellas in the doorway. (This was Oxford in summertime, after all.) I eavesdropped as the gallery assistant gushed about Ai Da. During our visit, we saw dozens of pictures and paintings. From these and some information sheets posted nearby, we learned that to produce a painting Ai Da's team follows this six-step process:

1. Ai Da's camera scans an object.
2. Ai Da's AI algorithm sends messages to a robotic arm.
3. Ai Da's robotic arm completes an abstract line drawing.
4. Human scientist Aidan Gomez, an Oxford University researcher, plots the coordinates of the drawing on the Cartesian plane.
5. Gomez feeds the drawing through a neural network.
6. Human painter Suzie Emery executes the final work.[21]

Retrieving my umbrella on the way out, I felt impressed, fascinated, and disappointed. The range and value of Ai Da's work impressed me. In 2018, a piece of Ai Da's artwork sold at Christie's for just under $500,000. The originality of Ai Da's creators, Aidan Meller and the robotics company Engineered Arts, fascinated me. The human ingenuity that went into posed photos of the tall, female, full-lipped Ai Da fascinated me, as Ai Da appeared poised, calm, and with eyes to camera. But this appearance also disappointed me, as "her" creators flirted with lazy, patriarchal tropes. And I was disappointed not to meet Ai Da herself.

Crook's face exuded disappointment. "It's not really advanced AI at all." (I think he wanted to meet Ai Da, too.) Of the above steps, only number 2 (and possibly number 5) involves artificial intelligence that is proprietary to Ai Da's creators. Much of the creativity lies in humans directing the work, either in creating the algorithms or in executing paintings. This is neither bad nor unimpressive, but Ai Da's intelligence is not existential.

Poet John Ashbery wrote, "The worse your art is the easier it is to talk about." Ai Da has this going for her: It's not easy to talk about her art. However impressive Ai Da's creators' achievements, there is a difference between an algorithm introducing randomization, stochastic noise, and other factors on the one hand and human artistic creativity and agency on the other. Ai Da's intelligence ignites interest and invites conversation, but differs in "kind," not "degree," from human intelligence. As curator and artistic director Hans Ulrich Obrist recently observed, computers cannot replace the artist.[22] To be human is to be distinctly intelligent.

HUMAN INTELLIGENCE v. ARTIFICIAL INTELLIGENCE

This is good as far as current AI goes, but what reasons do we have to think that humans will retain emotional and existential intelligence advantages over stand-alone[23] AI, as AI continues to improve? I think there are four reasons. First, narrow AI is doing worse than you may think at detecting even the basic emotions. Despite what some news sites would have you believe, human emotional intelligence dwarfs the artificial kind. No research has yet revealed consistent, specific fingerprints in the autonomic nervous system for different emotions.[24] It is rash to assume that narrow functional advances in affective computing today imply broad advances in artificial emotional intelligence tomorrow. There is a world of difference between AI helping crop scientists with logical problems and AI usurping our existential intelligences. The successors of Ai Da and AlphaGo won't usurp our intelligences any time soon. Ava is not on our doorstep.

Second, the standard psychological model of how we learn is learning by doing. We make mistakes and learn. Politicians will shrink from allowing AI to learn by mistakes in caring for our elderly relatives and in governing military departments. A dearth of meaningful opportunities to gain real-world experience will hamper AI's development.

Third, AIs may never gain the algorithmic efficiency needed to process emotions and existential questions sufficiently quickly or well. One scientist at DeepMind (developers of AlphaGo) described to me the resource-intensive nature of such processing. On commercial and computing grounds, he doubted whether algorithmic efficiency would keep pace with what would be needed if AI were ever to match human emotional and existential intelligence. Even if algorithmic efficiency grows quickly enough to keep pace, hardware efficiency may become the constraining factor, notwithstanding massive investment in AI-specific semiconductor chips.[25] Quantum computing will no doubt help, but AI may never surmount these obstacles.[26]

Fourth, in my view the argument stopper is consciousness. Even as AI continues to chalk up functional emotional advances, it will likely remain far from "having" those emotions in the way humans do. We have landed in philosophical territory here, territory that is sadly rarely showcased in media discussions about AI sentience.[27] As British philosopher Professor Colin McGinn wrote, discussing "consciousness can reduce even the most fastidious thinker to blabbering incoherence."[28] Computer scientist Stuart Russell starts in a similar place, observing that, "in the area of *consciousness,* we really do know nothing, so I'm going to say nothing," although he then proceeds to provide an eloquent and interesting fifteen-line footnote . . . about consciousness.[29] While I'm not going to say nothing about consciousness, I limit myself to the following comments.

The notion of qualia (singular "quale") is vital in discussions of consciousness. Qualia are individual instances of subjective, conscious experience, defined by philosopher and cognitive scientist Daniel Dennett:

> "Qualia" is an unfamiliar term for something that could not be more familiar to each of us: the ways things seem to us. As is so often the case with philosophical jargon, it is easier to give examples than to give a definition of the term. Look at a glass of milk at sunset; the way it looks to you—the particular, personal, subjective visual quality of the glass of milk is the quale of your visual experience at the moment. The way the milk tastes to you then is another, gustatory quale, and how it sounds to you as you swallow is an auditory quale; These various "properties of conscious experience" are prime examples of qualia.[30]

In my view, the risk of blabbering incoherence is greatest among those who claim that AI "has," or will soon "have," human emotional or existential intelligence. We just don't know what it means for an AI to experience qualia or to "have" emotions in the way we do.

For example, just now I tried asking Siri, "Why am I so stupid?" and received the impressively empathetic yet also quite misguided response: "I'm sorry to hear that. Talking to a friend or family member might help. If you want me to call or text someone, just ask." Siri uses advanced machine-learning technologies to function. However, few people would say that, by virtue of the above response, Siri has emotions or is conscious or intelligent in the same way as humans.

Russell puts it well: We do know that "machines are at a disadvantage when it comes to emotions: they cannot generate an internal simulation of an experience to see what emotional state it would engender."[31] Oceans of depth rest in qualia. Qualia look set to elude AI. At the threshold of an integrated AI-human future, the most inspiring, effective leaders take the time to dwell on how experiences feel and on the *way* something is, not just on *whether* something is.

THE IMPORTANCE OF STILLNESS

The moment when I realized how important stillness is for existential and emotional intelligence was in the spring of 2006.

Picture the scene. I'm sitting in a student study cubicle at St. John's College, Nottingham, United Kingdom, with a dilemma. After two years working part-time on a master's degree in theology, I'm finally ready to begin my 15,000-word dissertation. I'm sweating. I enjoyed thinking about philosophy, atheism, and Christian theology during my earlier, shorter, assignments. But I have always found writing difficult, since I'd focused in high school on math, science, and languages, ditching English essay subjects as soon as possible. Now, my dilemma was simple: How could I make this dissertation interesting? Over many weeks, I took the time to reflect, be still, and dwell on this question.

I decide to focus on Islam because the master's degree hasn't yet covered it. Only nine months have passed since the 7/7 bombings in London (four coordinated attacks in London on July 7, 2005), so I decide to make the dissertation topical by selecting the topic of jihad. Jihad is an Arabic word variously translated as "struggle," "effort," or "holy war." I'm still preoccupied with the question of how to make this dissertation *really* interesting—after all, the prospect of writing a long essay fills me with dread. In the quietness of my study cubicle, I consider what I uniquely bring that could be of value to others. Then an idea dawns: What if I meet the most extremist Muslim I can find and ask that person questions about jihad, and then also meet the most peace-loving Muslim I can find and ask the same questions?

In time, my dissertation became a reflection on my encounters with Musharraf Hussain and Anjem Choudary, experiences that led to some of the most fun and unnerving moments of my life. Hussain was vice chair of the imams and mosques committee of the Muslim Council of Britain, a group that received funding from the British government. Choudary led al-Muhajiroun, a proscribed Muslim organization. In 2002, he organized a conference in London called "The Magnificent 19," named for the nineteen hijackers behind 9/11.

Both interviews were fascinating, but the two and a half hours I spent with Choudary were spine-chilling. I managed to find Choudary's phone number. (Don't ask how.) I told him that I'd read about him in the press, that I was probably getting a biased perspective, and that I'd love to hear what he really wanted to say. He liked that.

Choudary agreed to meet, so I asked him where he'd be the next Tuesday. He laughed and said: "They keep taking my phone and raiding my flat. I don't know where I'm going to be tomorrow. Phone me on Tuesday and I'll tell you where to come."

I cleared my diary for Tuesday and called him. He told me to go to Paddington station in London and to call him when I was there. I was living in Oxford at the time, so I got on a train and went. When I called Choudary next, he told me to take the Tube[32] to Liverpool Street station and to call him when I was there. I complied and called him again from Liverpool Street station. Next, he told me to take the Tube to Ilford and call him when I arrived. Although I agreed again, I felt increasingly anxious and tense. I resolved that if Choudary suggested meeting anywhere that wasn't a public place, I'd turn around and go home. When I reached Ilford, I called Choudary.

"See that McDonald's opposite?" he said. "Go in there and I'll come in soon."

"How will I know who you are?" I asked.

Choudary laughed again. "You'll know."

I was a little surprised that someone committed to taking down the Western system and its symbols of capitalism would want to meet under the Golden Arches. Still, I went in, and a few minutes later a large man of Pakistani heritage sporting a lengthy beard and wearing a flowing robe came in. I ordered large fries, which we shared as we talked about jihad, interpreting scripture, and Choudary's aims in the United Kingdom.

At this point, I had no background in Arabic. But my interviews with Choudary and Hussain sparked my curiosity. I wanted to understand the Qur'an and associated debates on their own terms, for myself, so I enrolled in another master's degree, this time in classical and medieval Islamic history at the University of Oxford. There and in Damascus, I took courses in Islamic history and learned classical Arabic. I subsequently published a peer-reviewed academic article about these interviews that drew on history, theology, philosophy, and politics.[33]

Later, I went on to earn a doctorate in oriental studies at the University of Oxford, but my doctoral topic did not inspire me as much. If I'm honest, I pursued a doctorate partly in order to attain a badge of honor, the prefix "Dr." before my name. During my doctorate years, I sometimes rushed, not always pausing enough to consider deep nuances in source material. This scramble for achievement and my relatively simplistic approach impoverished my work. (Apologies to my inspiring and generous supervisor.) In some ways, the peer-reviewed article felt like my academic highlight in the humanities.

But the main thing I learned from my encounter with Choudary and my subsequent research is that cultivating stillness supercharged my performance, satisfaction, and impact. During my Nottingham degree, I wrestled with what mattered most to me and others. I cultivated curiosity. I took judicious decisions about what to study, based on how I could help society, not just to attain a degree. I sought to hold divergent perspectives and find larger patterns of agreement, disagreement, and commonalities among different groups. I felt I was somehow bringing my full being to what I was doing, more so than at any other point during my academic career. From quietness of heart, I brought my wisest, best self to my work and inspired others through related talks and a book. I got a distinction in my Nottingham master's degree, and I scraped through my doctorate.

Pinning your faith on AI to lead with existential or emotional intelligence is like walking in a wood at night with a compass and only staring at the dial. No matter how brilliant that compass may be, if you do not look up, you will walk into a tree.[34] Have you ever walked into a tree? Two years ago, I did exactly that and discovered just how painful it is. I was walking in a wood, joking around with my young daughters, not looking where I was going. I walked straight into a large oak tree, whacking the trunk with my knee, cutting my head on a branch and bruising my ego. A painful future beckons if we are not intelligent enough to look up and consider where we are going. Threshold leaders take the time to look up and nurture their own curiosity, wisdom, empathy, and intuition—both today and in the future when we discover enhanced ways of assimilating AI and humanity.

I agree with author and leadership adviser Liz Wiseman, who wrote that "There is more intelligence inside our organizations than we are using."[35] Leaders will flourish, and help others flourish, the more they embrace a richer, plural view of intelligence. This richer view is a gateway to the threshold, an essential leadership paradigm in an age of rapidly improving AI. We are not just upright apes living in dust with crude language and tools, all set for extinction. Human intelligence shimmers in a way artificial intelligence cannot.

THRESHOLD RESOURCES

RESOURCES FOR ANY LEADER

Resource 1. Liminal Questions
PURPOSE
Take small steps toward the threshold.

PROCESS
A brief reflection exercise. Step into the liminal space of threshold leadership by considering these observations and questions:

- In this breakneck world in which you must look busy or else, tune into the silences between the notes.

Ask yourself: Where can I cultivate stillness in my life, even for five minutes at a time?

· Usher in harmony and progress by holding divergent perspectives. *Ask yourself: Where in my life can I rely more on intuition? Where can I get more curious and empathetic about others, especially those most different from me?*

PAYOFF

Begin a journey toward AI-relevant leadership.

RESOURCES FOR LEADERS IN LARGE ORGANIZATIONS

Resource 2. Silent Aspirations

PURPOSE

Expand your team's AI aspirations.

PROCESS

An exploratory team exercise based in silence, intuition, and empathy. This exercise lasts three to six months and follows five steps:

· Lay the groundwork by gathering inputs from your team.
· Prepare to encourage your team to explore your shared aspirations at an off-site event.
· Explore your aspirations at a team off-site event.
· Decide on your three- to five-year aspirations.
· Specify the related nine-month aspirations.

I. LAY THE GROUNDWORK BY GATHERING INPUTS FROM YOUR TEAM

Collect views individually from your team on three topics:

A. Team purpose (eighteen-month to three-year time frame):
· How would you articulate our team's purpose?
· Why does this inspire you?
· What would you change about it?

B. Team strengths and passions:
- What distinctive capabilities or other strengths does this team have? This could include brands, networks, IP, privileged relationships, data science, predictive analytics, AI explainers, and enterprise-wide analytics.
- What AI or wider opportunities do you care about?

C. Stakeholder needs:
- What does your industry need? Where is the part of the market growing that is relevant to AI?
- In what parts of the market do customers most need help?
- Where is competition most intense?
- What parts of the market can AI credibly affect in our time frame and have the highest margin?
- What would most benefit society?

Get creative in how you collect views on A, B, and C. In advance of collecting their views, inspire your team members by encouraging them to find quiet, reflective space when compiling their answers. Guide them away from rushing their answers or squeezing in reflection time partway through a busy day. You might also consider collecting views as follows:

- Surveying team members using a web-form survey.
- Conducting one-on-one interviews with individual team members to delve more deeply into where they want to focus.
- Discussing the above topic at informal dinners.

II. PREPARE TO ENCOURAGE YOUR TEAM TO EXPLORE YOUR SHARED ASPIRATIONS AT AN OFF-SITE EVENT
Gather the inputs you received and reflect on them, with the aim of preparing for a rich, expansive, exploratory discussion about aspirations at a team meeting off-site. Ideally, you would perform this reflection with members of your team who have influence, regardless of seniority.

When preparing this discussion, avoid a purely logical exercise of categorizing or sorting. Also ask yourself what feels "hot" here. Put another way, what comments from your team seem to carry very strong emotional energy for them? Anchor the off-site around these emotionally resonant topics.

You will find one proposed way of structuring such an off-site in the Section III below. Find your own structure, using your intuition as you reflect on the inputs you received.

When planning the off-site, select an inspiring location that communicates "you matter" to your team. This will help them become more at ease with sharing what matters most for them.

III. EXPLORE YOUR ASPIRATIONS AT A TEAM OFF-SITE EVENT
SECTION I: VISION
Inspire team members to visualize the team's successful AI future. By doing this, you are deliberately cultivating stillness near the start of the off-site. Such a visualization exercise can focus on your team's or organization's inspiring future three to five years ahead. Many excellent resources exist online to design such an exercise.

SECTION II: SHARE INPUT
Share your off-site preparation with your team. Make sure to do this in an exploratory, invitational way. The mere fact of your having collected data in advance will, in the minds of some participants, risk putting you higher in the hierarchy than them. There is nothing wrong with being assertive with what you present. As well as this, create an environment where everyone knows that their thinking matters. The off-site will sparkle the more everyone knows they are designing an expansive journey together, rather than just receiving marching orders from one individual, which can be dull.

During this sharing and discussing, build in long periods of silence. As my colleague Brian Draper wrote to me, "The constant flood of words in life can block our deeper ways of being, and being together."

SECTION III: DEVELOP A DRAFT ASPIRATION
Together develop an initial hypothesis of your AI team aspirations. Encourage your team members to set the bar high with their aspirations. Five years is a long time in technology.

The threshold is a doorway into liminal space. Here are five questions or other inputs that may help your team articulate a shared purpose by stepping into this liminal space:

- Ask: "What it is that we can uniquely do that the AI world of tomorrow needs?" (Based on a similar question written by Trevor Waldock.)
- Say: "Let's move beyond naming a superficial 'purpose' rooted in status symbols or 'supposed to.' Rather, let's connect our thoughts about the future with our sense of what makes us fully alive."
- Ask: "Amid the buzz of technological progress, what makes you fully alive?"
- Quote German philosopher Friedrich Nietzsche, who said, "He who has a why to live can bear almost any how." Encourage discussion of the question "What is your 'why' in the area of AI?"
- Test the extent to which your emerging aspiration aligns with your broader corporate strategy. As Henke, Puri, and Saleh noted, companies that have scaled AI effectively are nearly four times more likely than others to align with such broader strategies.[36]

Try to keep outcomes provisional at this point. The more you explore the boundary between AI and humanity, the better.

IV. DECIDE ON YOUR THREE- TO FIVE-YEAR ASPIRATIONS
Whether at the off-site or later, together decide what your shared longer-term aspiration will be.

Allow space for each team member to do his or her own finest thinking about what you have developed. Encourage each person to say what they think, and not be swayed by the passion other team members may have shown for aspects of the aspiration. Modify the plan and reach consensus using your best decision-making tools. Ask, "When will we have achieved this?"

Also emphasize stillness in this part of the process. One way to do this before key discussions is to invite everyone to sit relaxed and with eyes closed if they feel comfortable. Some people prefer soft music rather than silence—this is fine.

V. SPECIFY THE RELATED NINE-MONTH ASPIRATIONS

A key step in bringing a long-term visualized aspiration to life is bringing it gradually from the future back to the present. This also should be done collaboratively, rather than by dictate.

For example, you might start by providing space for reflection and then discussion about the question "Given our long-term aspiration, what do you think your concrete nine-month objectives should be?" In whatever setting you are in, give each person time to pause. Listen to each other with fascination.

Before finally launching forward, come together again with your team and show the agreed aspiration statements. Have five minutes of silent reflection on the statements, while sitting or standing together, and then allow anyone to speak up. It's often in the silence that the best ideas emerge, and you will be surprised at how many useful comments emerge from such a silence.

From here, once you are aligned around your aspiration, all that remains is to deliver it! This might include the following steps:

- Decide how you will measure success.
- Set a thirty-day goal.
- Assess how ready you are to start as a team, and whose support you may need.
- Invite each team member to commit to a concrete small action in the next twenty-four to forty-eight hours.

PAYOFF

Progress into a more expansive AI-related team future.

CHAPTER TAKEAWAYS

- Human intelligence is plural, not just cognitive.
- The claimed dawn of imminent AI emotional mastery is false.
- AI is also a very long way from matching our existential intelligences on its own.
- At the threshold, cultivate space in your life to wrestle with who you are, who you are becoming, and what matters most.

2

CANCER IN
HIS SOPHOMORE YEAR

To attain knowledge, add something every day;
to attain wisdom, remove something every day.

—CHINESE SAYING

I recently read a story about a man visiting a carnival with his daughter, who asked for some candy floss. When the vendor handed it to her, he said, "Are you sure you can eat all of that by yourself, young lady?"

The little girl replied, "Sure, 'cause I'm bigger on the inside than I am on the outside."

This little girl knew that she had a greater capacity than what others could see. Although this only applied in her mind to candy floss, the same principle applies at the threshold.

TRUE AND FALSE SELF

You have universe-shaping capacity, which you will activate the more you embrace your "true self." The terms *true self* and *false self* were first introduced by the English pediatrician and psychoanalyst Donald Winnicott in 1960.[1]

Your true self is what makes you, you. When operating from this self, you feel alive and sense that you have a real, authentic identity. The true self prizes cooperation, harmony, and reverence for life. Your false self is your incomplete self, your defensive self, which you try to pass off as you. When operating from this false self, you may feel dead and empty in your thinking and being. Your false self wants to compete for everything, wants more of everything for itself, and thrives on conflict.[2] Your false self is inauthentic.

According to Dr. Robert Kegan and Dr. Lisa Lahey, every day "most people are doing a second job no one is paying for." This unpaid second job feeds the false self, with leaders "hiding their inadequacies, hiding their uncertainties, hiding their limitations."[3] In effect, they hide their authentic selves by covering up their weaknesses and managing other people's impressions of them, all of which take effort.

Uncovering your true self is a threshold competency essential for leaders in the Age of AI, with great payoffs for teams and organizations. The best way to access this competency is through stillness. Our focus in this chapter is on individual leaders, many of whom have moved from despair and inaction to hope-filled excellence by embracing their true selves and even realizing that these selves are evolving. As a result of doing this, one of my clients changed careers from business researcher to football club leader, another became a stay-at-home parent, and another went to work in Big Tech. In all these cases, the leaders felt they had stopped wearing a mask (of "being who they thought they were supposed to be") and had started embracing "who they really are" and who they were growing to be.[4]

For many, the journey from false self to true feels like finding summer through winter, or like singing their own song. It is a journey toward what matters most, which is so dearly needed as AI advances. In our world of monthly targets, social media pressures, and the next mortgage payment, it is not always easy to admit that what you have been driving at for years may not be what you really want to achieve. At stake, I think, is what it is to be fully human. At stake, also, is a successful AI future. Before coming to how threshold leaders uncover and embrace their true selves, we will first consider one crucial thing that leaders in the Age of AI need to start the journey: a humble approach to collaboration.

HUMBLE COLLABORATION

In 2020, Lieutenant Colonel Joseph Chapa finalized his doctorate in philosophy at the University of Oxford. With more than 1,400 pilot and instructor hours, Chapa is a thoughtful military leader serving in the US Air Force and establishing himself as a leading AI ethicist. Dr. Chapa returned to the United States in the summer of 2020 to serve on the Air Staff's Artificial Intelligence Cross-Functional Team at the Department of the Air Force. As of September 2022, he held the role of Department of the Air Force Chief Responsible AI Ethics Officer.

When I met Chapa to discuss AI, I started by asking him how humanity can thrive in such an age. Chapa threw me off center by musing on goats and ancient Greeks. "The Persian king Xerxes foiled the Spartans' defensive operation at Thermopylae by following a goat path around the mountain," Chapa said. "The Persians, unfamiliar with the territory, didn't initially know about this path, which bypassed the narrow pass that the Spartans defended." Chapa paused for effect. "But the goats knew better. By following the goat path, Xerxes was able to attack the Spartan position from the rear, and the rest, as they say, is history." As he related the story, Chapa flashed his winning grin. "Now, you and I don't know how to think like goats. It's not as though the goats set for themselves a logic puzzle and use if-then reasoning to solve it. And yet the combination of goat instinct, goat senses, and goat brains found the pass that the Persians could not initially find on their own. That's something like machine learning."

Chapa's point was that, if goat + human learning worked so well then, why can't machine + human learning work well now? This is where the true and false selves come in. There are times when we may want AI to come up with solutions that we couldn't have come up with easily or quickly, but our pride may block us from finding a good collaboration model. Threshold leaders jettison pride and defensiveness and open themselves—in the quietness of their hearts—to the true-self possibility that others' ideas may help us.

Personally, I find that pride clings. Arrogance regularly blocks my own efforts to find or express my true self, which can include openness to collaborating with AI in business and coaching. If you recognize yourself here, then be encouraged that even the smallest spark of desire to address pride shows a threshold trait. If nothing else, this spark indicates that you have something to work on. Use it to generate leadership momentum toward exploring the boundary with machines. What do leaders in the Age of AI need to start the journey toward embracing their true selves?

The openness and humility to recognize that a more collaborative form of human–machine leadership may be required.

There was another thing Chapa was keen to point out: "Keep in mind the idea that machine learning will spit out solutions that we couldn't expect. Sometimes it's the right answer. Sometimes it's not." The burden currently falls mainly on human leaders to tell the difference—mainly, but not entirely, as IBM's Project Debater demonstrates. Project Debater is "the first AI system that can debate humans on complex topics . . . to help people build persuasive arguments and make well-informed decisions."[5] Project Debater has debated human professional debaters and has the potential to support executive decision-making.[6]

Nevertheless, uncovering your true self includes working with paradox, being comfortable with risk, and showing the courage to reject an unwise solution, however logically compelling that solution may otherwise appear. Accessing the greater capacity of your true self therefore becomes vital in the Age of AI, because it's the best way to ensure that you're harnessing the best collective intelligence. Operating from their true selves, threshold leaders will be sufficiently centered to combine with AI, to spot valuable solutions even if they look strange, and as a result alleviate poverty, right injustices, and create planetary wealth.

AI and goats share this: They are unlike humans in many ways. Most goats have four legs, rectangular pupils, and four stomach chambers, whereas most humans have two legs, circular pupils, and one stomach chamber.[7] Recent AI tools typically have a few hundred neatly ordered neurons that get addressed one after the other, and these AI tools start tasks from scratch every time. On the other hand, a typical human brain has approximately one hundred billion neurons, does a lot of parallel processing (not in any particular order), and has a lot of structure wired into its highly evolved connectivity.[8] Like goats, maybe AI systems will be most useful where we work *with* them, or allow them, to do things differently to humans. Will you create silent space to explore who you are, enough to support a rich exploration of the boundary between humanity and AI?

In my experience, leaders access deep power when they cultivate stillness and silence. Sara Maitland wrote that silence is not "an absence of sound, but the presence of something which is not sound." In the final three sections of this chapter, we contemplate three vitalizing routes to stillness, appropriate for individual leaders seeking the threshold: stillness in motive, stillness in nature, and stillness in movement. Along the way, we will consider how these three routes stimulate leaders' journeys toward the true self.

STILLNESS IN MOTIVE

Ken Eagle wasn't the biggest or loudest young business leader I have ever met, but he had extremely powerful presence. I met him in 2019 at two leadership development events, coaching him during the second one. Nine years earlier in his sophomore year of university, aged just twenty, Ken experienced severe stomach pains and consulted a gastroenterologist, who told him, "It's bad and probably caused by alcohol and the copious amounts of pizza you are eating." Instead, the ultrasound revealed a large mass in lymph nodes outside his stomach.

What the gastroenterologist said next made Ken recoil: "I can't handle you from here. This is either an infection or cancer. You need a biopsy."

The biopsy came back positive. Ken had late stage 2 diffuse B cell non-Hodgkin's lymphoma, fortunately "only" localized. Ken immediately underwent six months of chemotherapy, shuttling between university and home. Said Ken: "Going through the experience, I interacted with the healthcare system on various fronts: an infusion center, a blood clinic, primary care, surgeons, a whole host of professionals in various care teams in different parts of the US. I saw patients of diverse backgrounds getting different levels of care. The care quality was very uneven."

Ken saw that the healthcare system dripped with potential for improvement. In the silence of the small hours, he made a professional choice to improve healthcare. Over time, this choice would narrow to wanting to improve the patient experience. On a personal level, he realized that life is short and resolved to avoid spending time on things that did not matter. In the face of death, Ken found life in quietly testing his motives.

His treatment went largely according to plan. Ken went into remission and, wonderfully, has since been cancer-free.

Fast-forward to mid-2019. Ken noticed that, in his work, he wasn't really improving the patient experience and was distracted by others' expectations of him. In recent years, Ken had taken various roles in healthcare. But in his own words, "I was getting closer to my goal of improving patients' experience but was still too far removed. I wasn't scratching my itch of improving the patient experience. During 2019, my wife started her own business in grocery and sustainability—really addressing a social need. Around this time, I considered leaving my goal behind and doing something different personally and professionally."

During a week I spent with him, Ken made a heartfelt public commitment to further investigate starting his own technology company in the personal healthcare experience space. This commitment represented a massive shift in outlook involving personal and professional change. How did Ken get to the point of making this shift?

At two incredibly busy points in his life, Ken was brave enough to find the tranquility to reflect on who he was, what drove him, and how he was connecting that with his vision. For a week, he devoted himself to exploring these things. This was threshold bravery.

Ken's coaches and team members asked him questions like, "What is it about you that creates the itch?" and "What motivates you?"

"As I mulled this over alongside my wife's new start-up," Ken said, "I realized that fear was blocking me, and I became more grounded in the passion I had carried for years."

I shared with Ken the following story:

> When Henry Nouwen, the acclaimed academic, gave it all up to become chaplain to a community of people with learning disabilities, no one there knew or cared for his illustrious reputation. "I might as well have been the janitor," he wrote. It rocked him. As he later reflected, he'd believed three classic "lies" about his identity— "I am what I do," "I am what others think of me," and "I am what I have."[9]

Ken built on this story during another exercise, in which he reflected on the passions, strengths, behaviors, roles, assumptions, and feelings that make him who he is. One silence at a time, Ken explored why he located these descriptors in himself and then identified themes that emerged.

What Ken did in this exercise was use stillness to connect to his motives. In a classic threshold move, he integrated his thinking with what he sensed of his being and refashioned his outlook. This led to a renewed sense of purpose, openness to collaboration, and energy for action. I feel privileged to have witnessed Ken share his public commitment; there was fire in his eyes. Here was a man who had uncovered his true self and acted. The more leaders in the Age of AI do this, the more they expand their capacity to thrive and serve others.

For some, silence is difficult to face. I think of Charlotte, a former professional dancer whom I coached when she had become a businesswoman. Charlotte shared with me two passions and a challenge. For as long as she could remember, her passions for creativity and compassion had brought her the most joy. Her challenge was her support team, the members of which didn't get on with one another and were badly underperforming. Over a period of months, Charlotte relaxed into being able to access inner silence.

Through a similar exercise to Ken's, Charlotte realized that the key to improving her team's performance was to be kind to herself and to tap into her and their creativity, even where she couldn't predict the outcomes. In other words, she uncovered her true self and thrived, which then helped her team. Once she viewed her team with compassion, their performance skyrocketed. In this case, Charlotte entered a liminal space, which fueled performance and joy. Quietly connecting to your deeply held motives will be crucial for you, too, as you lead in the Age of AI.

STILLNESS IN NATURE

Working with Brian Draper is different from working with other leadership gurus. Draper runs Echosounder, a consultancy that nurtures spiritual intelligence in leaders, often using nature walks. Draper also regularly contributes to *Thought for the Day* on BBC Radio 4. During the last fifteen years, Draper and I have served many organizations together. The beauty of working with Brian is that he moves at his own pace, creating spaces that still the soul.

Thus I was delighted when, in May 2019, Draper invited me on a retreat he was leading with Howard Green, a former science teacher and head teacher who works as an outdoor guide. At this time, my consulting work was gobbling up all my time and required me to shuttle between New

York, France, Austria, and Japan. On these high-octane trips, I'd tried and failed to create much meaningful rest. As I left for the retreat in question, billed as a "Spring Saunter," I felt weighed down.

We convened near Mottisfont Abbey, originally founded as an Augustinian priory in the year 1201. The name *Mottisfont* is Saxon and means "moot" or "meeting" around the font or spring. Ten of us formed a circle as Draper quoted Mary Oliver: "Attention is the beginning of devotion." I could already feel glimmers of peace return.

Draper led us to the Abbey's front lawn. We stood close to a deep well by a stream. Draper invited us to look down into the well, which has been there as long as there have been people, and probably long before. "Look at the water. It looks completely still, doesn't it?" asked Draper.

We all nodded, enjoying the calm.

Draper's next words jolted me to attention: "In fact, this well channels nine hundred liters of water per minute into that gushing stream." When I heard this, initially I didn't believe it. The water of Mottisfont well really does look completely still on the surface. But I looked again at the well and then looked to my right at a surging, frothing stream. There was no doubting it. Now I could see the water's power, its only source a serene spring that has always flowed.

For reasons I did not fully grasp, I lingered for what felt like hours. I recalled part of my own life motto, which is to be an artesian spring, surfacing life-giving water from deep wells of reflection, to serve others. Most of all, I observed productivity, flow, and stillness meet. I dwelt with this observation and felt refreshed.

Later that day, as we walked, Green pointed out the oldest tree in Hampshire, standing at around a thousand years old. We stood by the tree, still, imagining what it had witnessed during its life. Where countless other trees had fallen, this one stood. "Don't just do something, stand there," said Draper. There is something about being still that generates productive energy. Many leaders are exhausted and overwhelmed. But, like trees, unless we rest we do not renew, and we cannot sustain performance. This insight will be ever more fundamental as our AI-human future may be even more frenetic than is anticipated today.

The stillness we enjoyed reminded me of the following words of writer, speaker, and activist Parker J. Palmer:

The soul is like a wild animal—tough, resilient, savvy, self-sufficient, and yet exceedingly shy. If we want to see a wild animal, the last thing we should do is go crashing through the woods, shouting for the creature to come out. But if we are willing to walk quietly into the woods and sit silently for an hour or two at the base of a tree, the creature we are waiting for may well emerge, and out of the corner of an eye we will catch a glimpse of the precious wildness we seek.[10]

Like the thousand-year-old tree, Mottisfont well is at once still and productive. Appearing immobile, the well produces energy and the tree produces leaves. Threshold leaders are also still and productive. They are willing to sit at the base of a tree. They appreciate the indissoluble link between contemplation and action, dwelling on what matters most before rising to act. As the proverb goes, "Many words rush along like rivers in flood, but deep wisdom flows up from artesian springs."[11] In a noisy world where technology often interrupts us, silence amplifies the precious wildness of leadership and will help you explore who you are.

In my experience, the biggest barrier for people discovering who they really are and acting in alignment with it is that they think it will cost them. "All this stuff about being who you really are sounds great," the objection goes, "but I have three kids from two marriages, maintenance payments, a mortgage, car payments, and a health plan to pay for. It's just not realistic." An obstacle to this first threshold pathway of stillness is the relentless pursuit of more. Here's the uncomfortable truth: Yes, exploring who you really are may cost you. Perhaps it will cost financially, in the short or long term. Perhaps it will not. As Palmer and a host of mystics have pointed out, the path to flourishing is not up, but down. Like a tree's roots, down is essential. This really is often a difficult journey. But in my work with thousands of executives and other leaders, I have never met one who has regretted such a path.

A well, a tree, some woods. Almost whatever the cost, threshold leaders cherish the stillness of nature.

STILLNESS IN MOVEMENT

In recent years, together with Draper and others, I pioneered a way to help leaders cultivate stillness via moving and pausing: the labyrinth exercise. Labyrinth is a magnificent, centuries-old walking tradition. If you visit Chartres Cathedral, you will find the oldest surviving example of a labyrinth, from the twelfth century. It's a beautiful, peaceful, evocative, and intriguing space. A circular path on the floor, it looks like a maze, but in fact you can't get lost—it has one path in, which leads to a central space, and then one path out again.

More than ten years ago, Draper didn't need much persuasion to take the lead on updating labyrinth for today's working world. Since 2012, more than eight thousand business leaders have walked our modern, multisensory form of labyrinth. Standing against technological interruptions that trigger our emotions, our labyrinths help leaders wrestle with questions such as, "Why am I doing what I am doing?" "Why do I believe what I believe?" and "What am I fighting for?"

Labyrinth's enduring future power is existential. As we saw in the previous chapter, AIs can already mimic aspects of emotional intelligence such as empathy from a functional point of view. Yet the silence leaders create in modern labyrinth enables them to access bigger questions of life and leadership in resonant ways.

You might ask, what business does an ancient "walking meditation" have to do with the way you lead as AI improves? Whether or not you use a modern or ancient labyrinth, it offers you an extraordinary chance to reflect in a unique and dynamic way on the kind of soulful, centered leader you want to be in our AI future. In this way, you position yourself to capitalize financially, corporately, socially, emotionally, and spiritually in today's ever-changing world for the benefit of others. In labyrinth, movement and stillness meet in a gorgeous paradox.

Over the last few years, Draper and I have received countless emails, calls, and other personal outreaches from leaders testifying to labyrinth's lasting impact on their life. I highly recommend it as a gateway practice to threshold leadership. The threshold resource at the end of this chapter describes a simple way to access labyrinth.

YOUR SOUL ON FIRE

In my leadership work, I often meet people who feel paralyzed in their quest to reveal who they are. Sometimes their search for life's purpose, whether or not connected to technology, has become such a large and all-consuming task that the idea of succeeding in the future feels beyond them. At other times, an experience of trauma blocks them from making progress.

Uncovering true self can be hard. Perhaps you think you're not good enough as a person because of what others told you or tell you. This reminds me of the well-known story of a young student who wrote an economics paper on his vision for overnight mail and how he would achieve it. His teacher gave him a C grade and wrote, "Don't dream of things that can't happen." What did this student do? He left school and started Federal Express, the company that has one of the most ingenious logos in the world. He knew who he was and what he wanted.

Regardless of your background, you are a bright light in the universe, "a spark in the divine fire," in the words of author Jennifer Worth. And as Field Marshal Ferdinand Foch once said, "The most powerful weapon on earth is the human soul on fire." You can achieve what others think is impossible. Your wounds are real. But you can shape the world of tomorrow, no matter what arrows your teachers, parents, or colleagues have fired at you. The slenderest silent space may open the way for you to shift from fear toward love.

As AI forges ahead, old models of leadership won't do. Cultivating stillness is the powerful first path in a new threshold model of connecting thinking and being. By cultivating stillness in motive, nature, and movement, you will offer your most sparkling, powerful self to your colleagues, performing better as a result.

THRESHOLD RESOURCES

RESOURCES FOR ANY LEADER

Resource 3. The Art of Sitting
PURPOSE
Bring stillness to the heart of your leadership.

PROCESS

Digital clutter crowds out our space. An antidote to this is to sit. This may not sound revolutionary, but how often do you really just sit? Not sit and surf, not sit and read, not sit and talk. Just sit.

Try this: Choose a comfortable seat in your home or office and sit silently, waiting. Depending on your home or office environment, you may want to use noise-canceling headphones, too (with no music). It is not important whether you start with one minute of silence or one hour of silence. The important thing is that you feel at ease when you try.

As you wait, look out your window at the sky with a soft gaze. Try to clear your mind of clutter and seek stillness within you by focusing on your breathing. Also seek stillness outside you. Allow a natural silence to envelop your mind. This silence will help you let go of what holds you back, including exhaustion and emotional survival.

Imagine how much more effective your leadership can be as you connect your thinking to this kind of stillness at the heart of your being.

PAYOFF

Increase the effectiveness of your leadership by combining stillness and productivity.

Resource 4. Labyrinth

PURPOSE

Reflect on your own path through life and leadership.

PROCESS

This is a walking and pausing exercise.

One way to cultivate stillness in motive, nature, and movement simultaneously is to get outside and walk your own labyrinth. Find a labyrinth near you using an online locator such as https://labyrinthlocator.com/.

Alternatively, google "finger labyrinth" and print one out or have it on screen. Many people enjoy "walking" labyrinths by tracing their finger along or near the path. Remember to breathe deeply as you "walk" the path.

Labyrinths have three parts:

- An inward path, which leads to
  - A central space, and then
    - An outward path.

As you walk or trace your finger, consider the following.

On the inward path:
- Know that "this is a journey into the unknown, and yet it's a path on which you cannot get lost, so, this is not a gamble."[12]
- Ask yourself: Who has played a part in helping me to become the person I am? Since childhood, what have been my strengths? How has my work helped to shape the person I am becoming?
- Ask yourself: As I step onto a growth edge as a leader, what am I afraid of? What do I need to let go of?
- Breathe deeply.

As you come to the center of the labyrinth:
- Know that you come as you are.
- Orient yourself to where north is on the compass. Look in that direction and ask yourself: Who am I, really? Take your time. What are the things in my organization, my team, or other parts of my life that interfere with my "true north," my sense of direction? How can I remove some of those, so that the way becomes clearer?[13]
- Breathe deeply.
- Smile.

As you continue your journey on the outward path:
- Ask yourself: What am I fighting for as a leader?
- Ask yourself: What kind of difference can I make—now that I have sensed something deeper about who I am—in the world of tomorrow? In what way can I use my strengths, passions, and past experiences to lead our AI journey better?
- Breathe deeply.
- Smile.
- Breathe again.

Take time during and afterward to journal your thoughts.

To understand more about the transformative power of labyrinths, consult Brian Draper's excellent book *Labyrinth: Illuminating the Inner Path* (Lion Hudson, 2010).

PAYOFF

- Call forth wisdom by seeing the world with new eyes.
- Discover what from your past, your strengths, and your passions best equips you to navigate an increasingly integrated human-AI world.

CHAPTER TAKEAWAYS

- Uncover, embrace, and work with your true self.
- Humbly invite a more collaborative form of human–machine leadership. This requires ease with openness, risk, and paradox.
- Cultivate stillness in motive, nature, and movement.

3

"MAN, YOU'RE GONNA BE RICH!"

Life is not a matter of creating a special name for ourselves, but of uncovering the name we have always had.

—RICHARD ROHR

It all started so well. It's a crisp spring afternoon in 2015, and I'm drinking tea in London with my friend and former colleague Lord Nat Wei. We're discussing entrepreneurialism and how the world can benefit from coaching in a world of wearables and AI. Suddenly, Nat suggests an idea: "What if you put wearables and coaching together and create an app for people to track and improve their own performance?"

I could sense the power in the idea. It felt like its time had come. Within weeks, I'd launched a tech start-up called Coachify, and I was the CEO. Over the following two years, I learned more about leadership than I could ever have imagined.

Fast-forward to the summer of 2016. We'd piloted version two of our beta product to delighted customer reviews. We had interested buyers, a passionate sales staff, a product pipeline including Ainsley (our own AI coaching bot), and externally commissioned iPhone and Apple Watch products. So far, so good, although dark clouds loomed on the horizon.

It's worthwhile dwelling on three factors that helped us achieve this initial success. First, we knew who we were as a team. One of the things I did well was to get the right AI talent in the right roles. Of course, talent in an AI world includes artificial as well as human intelligences. Before appointing anyone (human), I mapped candidates on a matrix of skill and will to assess what role would best suit them and gauged the extent to which we shared common values about building AI. Alongside this, customers loved how we understood their own needs and hopes. Second, our teams united around a compelling, shared purpose. Two experienced sales executives even offered unpaid time in order to participate. Third, we challenged deep-seated ways of thinking about coaching, as we were clear on our values. We may summarize these three factors as awareness, purpose, and values. We had the understanding to know who we and others were, the inspiration to know what we wanted, and the wisdom to know how we wanted to journey through life.

Indeed, the questions that customers and investors most often asked me during my time as Coachify CEO fell into the same three categories:

- Awareness: Who is your team? Who are your customers? (Or: Who are you?)
- Purpose: What do you want your organization to achieve and why? (Or: What do you want and why?)
- Ethics and Values: How will you deal with data privacy or other ethical concerns? (Or: What values guide you? How will you travel?)

This chapter explores how threshold leaders can create spaces to access these three factors in their teams and organizations as technology advances. Before exploring each factor in turn, I offer some general comments about the importance of stillness for organizations in the Age of AI.

THE POINT OF NO RETURN

Make no mistake, AI is accelerating fast. According to research published in March 2022, artificial intelligence will contribute an additional $16.5 trillion of GDP (gross domestic product) to the world economy by the year 2030.[1] Four factors are supercharging AI's growth: exponential advances in computing speed, theoretical breakthroughs, trillions of dollars, and vast amounts of data.[2] This is a potent combination.

The downside of getting our future relationship with AI wrong is enormous. We already know that AI has been used to manipulate democratic processes, violate privacy, attack national security, and cause reputational and financial damage for organizations.[3] Automated, poorly understood credit ratings could completely cut people off from getting business loans and from access to education, and we won't even know until three decades have passed and a whole generation has been stratified.

People often ask me, could it really go that wrong? I'm an optimist but, yes, of course it could. "I'm very close to the cutting edge in AI," noted Tesla and SpaceX CEO Elon Musk, "and it scares the hell out of me."

In 2021, Microsoft CEO Brad Smith warned that "if we're not careful, George Orwell's *1984* could come to pass by 2024."[4]

Not only that, maybe we will one day meet HAL, the fictional AI machine in Arthur C. Clarke's *Space Odyssey* series. In the film *2001: A Space Odyssey*, Dave Bowman asks HAL (Heuristically programmed ALgorithmic computer) to open the pod doors on his spacecraft as a precursor to disconnecting HAL. Somehow smelling a rat, HAL famously responded, "I'm sorry, Dave, I'm afraid I can't do that." How long will humans retain the agency to stop hell on earth from coming to pass?

Automation could wipe out hundreds of millions of jobs. I am unconvinced that AI will remove jobs quickly or completely, but those negative effects that we do see will likely be unequally distributed. Consider the aggregate picture first. Dr. Carl Benedikt Frey, one of the Oxford academics behind the finding that 47 percent of American jobs are at high risk of automation by the mid-2030s, turns out to be no prophet of doom.[5]

New jobs will also arise as AI develops.[6] As author and businessman Kai-Fu Lee put it, "Explain to someone in the 1950s what a 'life coach' was and they'd probably think you were goofy."[7]

One historical data point is relevant to this part of my argument. In the 1810s, 84 percent of the US workforce worked in agriculture.[8] Today, 3 percent do.[9] There was no period in between where the transition caused unemployment to soar to anywhere near 47 percent.[10] Taking a long perspective, we may see that—at an aggregate level—new jobs may replace a significant number of lost jobs.[11]

But we have also seen labor market changes exacerbate inequalities. By definition, diverse populations cannot adjust equally well to change.[12]

Classics and history professor Walter Scheidel argues that the only things that have reliably reduced inequality in the past are plagues, revolutions, massive wars, and collapsed states.[13] AI-related wealth does look set to concentrate further among both nations and corporations.[14] We have little reason, then, to think that automation will affect workers equally. AI does seem likely to remove some jobs in an unequally distributed way, a considerable downside.

AI-enabled surveillance, sex bots, and blackmail schemes could further destabilize society, which will affect the context in which organizations operate. Recall E. O. Wilson's challenge about paleolithic emotions, medieval institutions, and accelerating godlike technology (see the chapter, Meet The Threshold). As AI accelerates, this challenge really comes home to roost.[15]

With every passing year, lack of competence is becoming less of an issue for AI. In 2019, DeepMind's software beat 99.8 percent of human competitors in the fiendishly difficult strategy game StarCraft II, achieving the elite grandmaster level. In 2022, the AI agent Gran Turismo Sophy won a head-to-head competition against four of the world's best Gran Turismo drivers. To succeed, this AI not only achieved the fastest route around the track, but had to balance this with the difficult problem of interacting smoothly with human players who are often unpredictable.[16] This is quite some progress from an early English-to-Russian translation program, which "is said to have translated the sentence, 'The spirit was willing but the flesh was weak' to 'the vodka was good but the meat was bad.'"[17]

It is not inconceivable that accelerating godlike AI will usher in an apocalypse. Governments and non-state actors are investing heavily in autonomous weapons systems—think nuclear weapons on speed.[18] And philosophers have imagined more ways that AI could wipe us out than there are paper clips on the planet.[19] As our universe becomes more AI-infused, and quickly, it is not hard to imagine a future where the scariness, brilliance, creepiness, and scale of AI increases forces of social disorder and even collapse.

Without a different course, it seems we are getting closer to a point of no return. Complicating matters is that AI will be a reflection of humanity, especially if it's trained by vast amounts of data generated using humans. As Ray Kurzweil put it, "nonbiological man" is closer than you think.[20] AI is formless, invisible, and—according to some—will soon echo the divine: omniscient and omnipotent.

STILLNESS FOR ORGANIZATIONS

Of course, Wilson wrote more about humanity than AI or God, but our trajectory with AI does put us on a precipice. For many, these observations about the downsides or trajectory of AI cause fear to rise. If you feel fearful, you're not alone, and these are natural feelings. Consider the words of Richard Rohr, the founder of the Center for Action and Contemplation. Writing metaphorically, Rohr wrote: "If it is our temperament to seek security, we will run back to the old room that we have already constructed. If it is our temperament to take risks, we will quickly run to a new room of our own making and liking. Hardly anyone wants to stay on the threshold without answers."[21] As AI evolves, how do you feel about not having all the answers?

One of the best actions to avoid a precipice is to take your bearings well. In an organizational context, this includes cultivating stillness in your workplace. In other words, this means putting yourself in a liminal space and staying there long enough to learn something. Rohr explains why this matters: "Liminality keeps us in an ongoing state of shadowboxing instead of ego-confirmation, struggling with the hidden side of things, and calling so called normalcy into creative question." As AI evolves, effective leaders will access this state of shadowboxing, regularly calling so-called normalcy into question, as normalcy will keep changing. As you regularly put yourself in liminal space, you best position yourself to inspire others, create well, and avoid destruction. Let's consider what this looks like for organizational leaders.

At their best, AI-integrated organizations can be powerful resources to amplify the benefit of societal, national, and global goods. Our recent history illustrates this.

The chaotic year of 2020 brought unprecedented progress in the field of AI and gave us a new sense of what is possible. In the weeks after COVID-19 emerged globally, even organizations with limited analytics experience set up effective AI-enabled crisis-response solutions with astounding speed. For example, it took only twelve months to deliver two novel mRNA vaccines, a process that would have taken a decade under usual circumstances. Machine-learning systems helped researchers quickly understand the novel coronavirus and its structure and predict which of its components would most likely provoke an immune response.[22] When the next pandemic strikes, AI will be an even more indispensable part of a medical researcher's tool kit.

There was a stillness at the heart of one of the most effective leadership responses to COVID-19. Even at the height of pandemic panic, AstraZeneca promised to offer vaccines at cost price. Cynics claimed that AstraZeneca did this to get a foothold in the lucrative vaccine market for the longer term, and the cynics may have a point. This doesn't change the fact that, in the shorter term, AstraZeneca deliberately took a profit-destroying path to develop a vaccine that would enable hundreds of millions of people living in the poorest and most fragile regions of the world to get vaccinated.[23]

What created the conditions for AstraZeneca to take such beneficial action? Just two years before the pandemic, it launched the AZ2025 campaign, an initiative designed to shape AstraZeneca's path toward sustainable success. Two key planks of the campaign were "listen" and "think." Through roadshows and other events, participants paused, considered, listened, and only then spoke.[24] In early 2020, AstraZeneca's leaders showed that stillness fueled their decision-making, as they were centered enough to peg the pandemic early as a species-wide challenge. At this point, according to his own recollection, CEO Pascal Soriot immediately prioritized global health above price maximization.[25]

Be in no doubt: Prioritizing anything above price or value maximization is hard for major corporations. AstraZeneca, for example, was struggling in 2012 when Soriot took over as CEO. Ten years later in July 2022, AstraZeneca was the largest company in the FTSE. Market capitalization growth has been a large part of AstraZeneca's story.[26] Given this reality, AstraZeneca's focus on the needs of middle- and low-income countries was especially laudable, and it only happened as a result of Soriot's personal drive and leadership.[27]

Led by Soriot, AstraZeneca showed genuine interest in others, articulated a clear "why," and guided its teams toward opportunities to collaborate, even with competitors. Its initial vaccine success stored up political and public-relations problems. When manufacturing couldn't keep up with demand, French president Emmanuel Macron described AstraZeneca's vaccine as "quasi-ineffective" before volunteering to have the jab himself.[28] Nonetheless, AstraZeneca led from awareness, purpose, and Rooted Values (more on this at the end of this chapter), averting millions of infections as a result. I view these as threshold-level leadership responses.

There are few better ways to influence the future than by creating stillness at the heart of organizations.

This brings me to one of the most exciting things I have learned in my whole career: Whatever leaders think they are doing when they lead, they are revealing something of their being. Some executives chase dollars at all costs, revealing that they value earnings, personal success, or fame above all else. Others prize multiple awarenesses, including about society, ethics, and a diversity of resources. This learning matters in an age where AI increasingly accomplishes narrow goals. Leaders who take refuge in purely technical or managerial skills and who fail to inspire cultures where others live from their true selves will soon find themselves irrelevant and alone.

Reflecting on this, I realized that the most successful organizational leaders connect their thinking and their being and help others do the same. This means creating the space for others to explore who they are, what they want, and how they will travel there. Researchers have discerned this approach to be an essential, underlying paradigm built into the structure of thriving organizations. For example, over nearly two decades, consultants Scott Keller and Bill Schaninger led a vast, longitudinal, multidimensional research effort into organizational change. They found that at the core of leading change in a centered way is being a centered leader, not just thinking well.[29] Integrating thinking and being through stillness empowers organizations to achieve much more than individuals could do alone.

AWARENESS

Back at Coachify, just as things were going well, the problems started to mount. By late summer 2016, being CEO was a full-time job with increasing demands. The hours were exacerbated by repeated and elongated 6:00 a.m. phone calls with the product development team. Although my wife and I were funding the start-up, I realized that Coachify needed more, so I hit the international fundraising trail, networking with a succession of high-net-worth individuals, accelerators, and other investors. I was also delivering executive coaching on the side to bring in some funds. I started to burn out, although I didn't notice it at the time. I felt increasingly afraid of failure and started griping at suppliers and my family. I felt disconnected from my best self. Joy slowly seeped away.

In early October, I temporarily axed app development and sales, switching my focus to fundraising. Within weeks, I'd assembled two separate consortia, each offering the required seed funding. Then a new investor, whom I'll call Oliver, offered to join a consortium. After our fifth meeting, he offered multiple millions of dollars (way above my initial request), in return for control of the business and my moving abroad to base Coachify nearer to him.

My response to Oliver's offer was telling. If I'd known that leading Coachify was deeply connected to who I was and what I wanted to bring to the world, I would have said: "I'm glad you see value in my business. No, I'm not giving you control of Coachify at this early stage. Let's talk about how you could get involved."

Instead, I replied: "Hmm, yes, that's interesting. Let's talk about how that could work." After two more discussions, two days before Christmas, I received a one-line email from him: "I'm withdrawing my interest." I haven't heard from him since. To this day, I don't know why he walked away, although I suspect he discerned something of what I later discovered.

The next day, on Christmas Eve, I caught the flu. My aching body was telling me something. As I collapsed in bed, a question suddenly popped into my head: *What if I don't do this anymore? What if I hand over the leadership of Coachify?* I felt sparks of joy and relief. In the coming weeks, I decided to turn down the remaining investment offers and step away from the business. When I tell this story to start-up executives or business leaders, the most frequent response I receive is, "Wow, moving on must have taken such courage!" I find this response odd, because it took no courage at all. Maybe others assume that my natural path was to want to lead a unicorn.[30] I remember a friend to whom I explained my idea in 2016 and who replied, wide-eyed and shouting, with hands in the air, "Wow! Man, you're gonna be rich!" But my goal wasn't to make money.

What was really going on? In those final months, I failed to cultivate the stillness required to help Coachify stay true to what it was. My false self came to the fore. It's normal for start-ups to pivot their strategy.

(Sometimes "pivot" is start-up slang for "our previous concept did not work," but that's another story.) During internal strategic discussions and negotiations with potential investors, Coachify had gradually pivoted toward being a data company. Coachify's financial value lay in the high-quality data we could harvest from business executives using our app. But I didn't want to lead this kind of company, and my team wasn't inspired by such a pivot.

Stillness helped me create awareness that liberated my team and me. How did this happen? My bedridden, calm reflection initially destabilized me as I realized that I might move on. Next, using threshold resources such as those in chapters 1 and 2, I (re-)discovered something I didn't expect—a golden thread that had been in my life and career all along: to help people flourish and to help organizations sustain top performance by working directly with them. As I clambered back to life, I quietly took refuge in poetry. These words felt resonant for me:

> We shall not cease from exploration
> And the end of all our exploring
> Will be to arrive where we started
> And know the place for the first time.
>
> —T. S. ELIOT

As poet David Whyte points out, poetry may not be your route to stillness. It is not always mine. "Having two children, my assistant finds the concentrated busyness of her mornings in my study-office a kind of bliss," Whyte wrote. "Quiet and contemplation in the office do not necessarily have to be in the form of a special room for silence and meditation; it could equally be in the form of a company culture that encourages people to admit they do not always have an answer."[31]

I also used stillness to find gifts in the upset I had suffered in the weeks from Christmas onward. The upset was not so much from Oliver walking away or from pain at falling ill. My upset was from sensing that I might soon let go of something to which I had devoted myself intensely for nearly two years. Although the idea of letting go gave me some joy, it initially unsettled me deeply as I couldn't see an alternative. One of the gifts I found as I emerged from the fear, negativity, and near-burnout of my Coachify experience was that, bruised, I stood on the threshold of the most successful and satisfying years of my career.

Who knows what upsets AI may deliver, including where it may challenge your leadership edge as the human boundary with AI blurs? One thing I do know: Leaders who find places of stillness will be more likely to progress into fruitful forms of humanity as they nurture self-compassion.

How did I create stillness to help my team? I sought Coachify's true self, not just my own individual true self. In a collective context, true self means what makes your team a team, and what makes your organization an organization. A team's false self competes too aggressively, whereas a team's true self collaborates, even across silos and with competitors. So, first, I asked questions that created space for colleagues to discover and reveal their needs and hopes. For example:

- How does your job relate to how you see yourselves?
- What if you knew that you are not what others think of you?
- How can we collaborate with our customers, suppliers, and competitors?

Second, I encouraged my colleagues to cultivate stillness in the three ways highlighted in the previous chapter. These actions opened the way to rich discussions about how we bring who we are to our shared endeavor. In short, I created a culture of stillness. Ultimately, my awareness of true self deviated from that of Coachify and its teams, so I stepped away. I'm heartened to see leaders in other organizations take forward similar missions, much more effectively than I would have, had I led from "false self" awareness.

PURPOSE

The second area where team leaders and organizational leaders can cultivate stillness is in relation to purpose. One of the few things AI leaders I have advised all agree on when discussing AI is that purpose matters. Here is a selection of what these leaders have said to me during the last two years about purpose:

> Are we a product company or a purpose company? We need to be the latter if we are to address the most critical and urgent issues facing our planet in this "decade of action."
>
> Technology is only a means to an end. It is not the endgame. Let's consider why we are doing this.
>
> The most important question your AI strategy needs to address is: What's the purpose of deploying automation?

Leaders of the largest organizations share these sentiments. Two examples will suffice. Real estate and retail conglomerate Majid Al Futtaim (MAF) has invested hundreds of millions of dollars to build its AI-driven consumer-sales strategy.[32] MAF's CEO, Alain Bejjani, is no stranger to world-shaping issues. In 2019 he was one of the eight cochairs of the World Economic Forum on the Middle East and North Africa. In October of the same year, Bejjani said: "Tomorrow is not going to be like yesterday. Tomorrow is going to be more human, smarter, and purposeful, and all of this can happen with better technology and integration with AI to transform the customer experience."[33] Leaders such as Bejjani realize that, to succeed in the Age of AI, we will need more than technology or cognition alone. We will need something more purposeful, more fully human.[34]

At least twice since 2015, Coca-Cola has taken the time to seek out and operate from a renewed purpose as an organization. These efforts resulted in putting AI at the heart of its growth. In 2017, Greg Chambers, global director of digital innovation, said: "AI is the foundation for everything we do. We create intelligent experiences. AI is the kernel that powers that experience."[35] Coca-Cola then harnessed AI in its product development, local product mixes, and new manufacturing methods such as augmented reality and social data mining. This led to thoroughgoing digital and cultural transformation efforts, accelerated during the pandemic. This is an unfinished story, but early signs of how Coca-Cola is putting purpose into practice are promising.[36]

In the words of Antoine de Saint-Exupéry, "If you want to build a ship, don't drum up the men to gather wood, divide the work and give orders.

Instead, teach them to yearn for the vast and endless sea." Whatever line of work you are in—education, retail, politics, or elsewhere—teach others to yearn for the vast and endless sea of possibilities that are relevant to you.

We know from several research studies that purpose-driven organizations improve performance and satisfaction.[37] AI can also speed up this improvement, as IBM's Tina Naser and colleagues wrote in their article "Accelerating the Journey to Purpose Through AI."[38] Two resources that will help you define inspiring goals are Labyrinth and Silent Aspirations. I encourage you to use these resources, and those at the end of this chapter, with your team. These resources will help you access calm, strength, and courage, which are distinctly human qualities that machines can't match, especially when fueled by stillness.

ETHICS AND VALUES

So far, we have explored silence and stillness as ways of centering your team and organization on who they are and what they want. Silence and stillness can also help you discern the ethics and values that will guide you in the Age of AI. Cambridge University researcher Dr. Alexa Hagerty put it well: "I think we are beginning to realize we are not really 'users' of technology, we are citizens in a world being deeply shaped by technology."[39]

When Lieutenant Colonel Joseph Chapa (whom we met in chapter 2) and I recently spoke, we focused on the topic of responsibility, as described in the first question above. "Animating this principle of responsibility is difficult," said Chapa. When asked about who he thought was responsible for what happened to Heidi Waterhouse, Chapa looked pensive. "It's not clear. That's part of the problem. In a military context, we can't hold the machine responsible because, well, it's just a machine. And maybe we can't hold the leader responsible because they might not know enough about what's going on under the hood of the algorithm to be responsible for the use of that algorithm. Maybe we should hold the developer responsible, but developers often lack broader context for how their work is being used, especially if the developer is a contractor. This is a difficult challenge."

Pressing on regardless of this challenge often backfires. Chapa noted that earlier AI leadership attempts suffered from a poverty of reflection. "When developing AI systems in the 2000s and 2010s, we rushed ahead to patch up our ethical approach to historic machine-learning systems," he said. Turning to solutions, Chapa shook his head slowly. "When we employ AI in the future, we will have to decide in advance who is going to be responsible for each element in the system, and how are we going to execute that responsibility. This takes time. We need to slow down and proactively decide about responsibilities in relation to highly advanced AI systems that have not yet been fully developed. And this is even before we get to the issue of biased data, which produces biased results. Many machine-learning systems are not equitable, and we don't know who's responsible when problems arise."

I am not so naïve as to claim that humans will never be in danger from machines. Technology can get in the way, and we sometimes see a poverty of leadership responses to pressing challenges. But history is full of examples of humanity rising to the challenge. One example is medical ethics, which developed from almost nowhere since the 1960s, including in pharmaceutical companies, and avoided many potential pitfalls such as unsafe research trials.

What are the ethical guardrails you will set up to prevent some humans being seen as more relevant than others? As Kai-Fu Lee wrote in his book *AI Superpowers*, "Building societies that thrive in the age of AI will require substantial changes to our economy but also a shift in culture and values."[40] Discerning what values to emphasize is not a simple task. Willing advisers flood this discussion, including leadership consultants, spiritual advisers, ethicists, and AI organizations.[41] My contribution is not to propose a definitive list of values, but rather to suggest that leaders who cultivate stillness in the hearts of their teams and organizations are the ones who will best navigate ethical discord.

As Rainer Maria Rilke wrote, "I am the rest between two notes which are always somehow in discord" (translated by Robert Bly). At the threshold, you are that rest and you inspire your organization to be that rest. Our fast-developing AI paths need not end in catastrophe. In a technologically advanced context, organizational serenity may be a godsend.

THRESHOLD RESOURCES

RESOURCES FOR LEADERS IN LARGE ORGANIZATIONS

Resource 5. Rooted Values

PURPOSE

Cultivate stillness in how your teams navigate tricky issues around ethics and values.

PROCESS

This resource involves a months-long, system-wide initiative.

Many organizations have found the following eight steps useful in articulating their values in a way that positively affects their culture and behaviors. The steps ideally occur in parallel with, or after, the organization as it develops its mission and vision.

- Formation: Select a core "values" team to lead the following process. Select this team inclusively and diversely. As Professor Ruha Benjamin said, "**Focus on building the right team before you start building AI systems**. Diversity needs to start from the groundwork that happens before the foundation is poured."[42]
- Ideation: Have multiple stakeholders generate long lists of values, using focus groups and other creative informal means. Challenge participants with questions such as: What balance between supportive and challenging leadership do you value? What is more important to you: data privacy or utility (and why)?
  - Take your time during ideation. Allow plenty of time for silent reflection.
  - If the discussion focuses on purely financial aspects of value, introduce wider considerations such as humanitarian, societal, planetary, and/or universal value. You might also direct people to Stuart Russell's book *Human Compatible*. In this book, Russell articulates a route to what he calls "provably beneficial" AI. He advocates building altruism and humility into machines.

- Grouping: Group the long lists into categories, arriving at a first hypothesis list of around four to six values, each with sub-points as needed. Aim to make these values as universal as possible for your organization
- Specifying: Assign pairs of positive and negative behaviors to each value. Here, you articulate "what we do" and "what we do not do" as a result of each value. I have found this to be a critical step if your values are to be widely adopted long term. Without this step, values feel vague. Vary the behaviors by sector, function, or another unit as required.
- Socializing: Return to many of your stakeholders and test the draft values. Allow them plenty of time and space to reflect on what you shared and to recall stories about the values you shared, remembering that they were not initially involved in developing these draft values.
- Planning: Plan how you will embed these values in every meeting, review, training, and in other systems and processes.
- Agreeing: Formally adopt the values through mechanisms that matter for your organization, such as performance ratings, processes, and remuneration decisions.
- Implementing: In addition to the above plan, capture symbolic opportunities to reinforce the values, such as website assets, off-sites, or major town hall events. In one organization I helped, when the senior leader moved to a different organization, she physically handed over the values, embossed on placards, to the new senior leader at a public handover meeting.

Here are two practical ways that threshold leaders use stillness to fuel a values exercise:

1. Have each team member write down qualities that they appreciate about one another. Ensure that the writing time is spent in silence. "Qualities" refer to who the person is, not what the person does. The format you might encourage is, "One thing I appreciate about you is . . ." This exercise could be done in a team meeting or asynchronously, for example, via instant message or handwritten

postcard. Appreciation costs nothing, takes very little time to do, and works.[43] Qualities that people admire in others are often closely related to values that they themselves cherish.

2. Establish a regular cadence for your team to sit in silent reflection, contemplating topics such as one of the six ethical AI questions in Appendix 3. Here's a possible structure for this exercise:
   - Introduce your relevant ethical topic briefly.
   - Set a timer for five minutes of "stillness" or "soulfulness," perhaps accompanied by soft music that lacks a beat (for example, Ólafur Arnalds's beautiful song "Fyrsta"). During this time, encourage everyone to sit relaxed and with their eyes closed or gazing softly ahead.
   - After the five minutes, ask each person how he or she feels.
   - Then gently start the discussion, encouraging everyone to maintain an approximate equality of speaking time among the group.

Payoff
- Lead purposefully into your AI future.
- Limit reputational damage.

CHAPTER TAKEAWAYS
- AI is accelerating fast and could go badly wrong.
- One of the most important antidotes to this is cultivating stillness in organizations.
- Threshold leaders will do this by helping others explore their true selves, purpose, ethics, and values.
- Create space for your teams and organizations to discover who they are in an AI world, what they want, and what will guide them.

THINKING INDEPENDENTLY

4

CODING ERROR

If I'm going to sing like someone else,
then I don't need to sing at all.

—BILLIE HOLIDAY

In 2018, I binge-watched season four of the dystopian science fiction Netflix series *Black Mirror* created by Charlie Brooker. The final episode of the fourth series is called "Black Museum." According to the episode description, "On a dusty stretch of highway, a traveler stumbles across a museum that boasts rare criminal artefacts—and a disturbing main attraction." The owner and proprietor of the Black Museum, Rolo, explains the sad story behind one of the exhibits, doing so with a twist that illustrates what many AI writings and TV programs wrongly assume about who we are and what we may do.

The story goes like this. Jack and Carrie are madly in love and have a son, Parker. Carrie gets run over by a truck and ends up in a coma in an advanced hospital where they offer Jack the chance for an implant that digitally extracts a patient's consciousness and rehouses it in a host brain. During the show, Carrie's consciousness is taken from her and transplanted into Jack's head. Jack eventually finds a new partner, Emily, who becomes understandably exasperated by "Carrie" being in Jack's head.

A key scene in this story is where Rolo offers a solution to the problem: The permanent deletion of Carrie's consciousness and the prompts that her thinking provided Jack. And then the following conversation happens:

> **Jack:** But that would be killing her, legally.
> **Rolo:** But not ethically.
> **Emily:** Please, she's just some leftover code in your head. It'll be like . . . like deleting an email.

I won't spoil what happens. But this episode depicts Carrie-inside-Jack's-head as a fully human Carrie, with emotions and access to wisdom, intuition, and creativity. In Emily's mind, Carrie is nothing more than computer code. Carrie's thinking processes are presented as entirely reducible to equations and algorithms. This assumption goes unchallenged by Jack and Rolo in the scene.

In Path II of this book, we explore how independent thinkers hold the future of the universe in their hands, or rather their minds. We are not just computer code or a biological boot loader for digital Superintelligence.[1] "The quality of everything we do is driven by the quality of independent thinking we do first," noted author Nancy Kline. I agree with Kline, especially in a context where AI increasingly integrates into our lives. The boundary between humanity and machines is, after all, already blurred. Many people spend just a few minutes per day more than one meter away from their smartphone.[2] AI pervades smartphone software. We are not very far from having AI-enabled implants. AI is around us, almost within us, already challenging or otherwise interacting with our independent thinking.

In this chapter, we'll investigate why human independent thinking matters as AI improves. We will consider three interruptive thinking disrupters, before starting to explore what independent thinking looks like practically.

INTERRUPTIVE TECHNOLOGICAL FORCES

Few things crush progress more than interruption. Kline often calls interruption "an assault on our thinking process." More than this, technology increasingly "colonizes our attentional space" and "depletes our cognitive autonomy" by interrupting us.[3] Take for example the phone notifications that trigger us or the marketing ads that create desire and amplify pain. Recently, I turned off every notification on my phone. I have felt more centered and productive ever since. But it took considerable willpower and a whole hour to complete the task. Not without a fight can we escape even a little digital colonization of our thinking.

Kline and her colleagues have identified fourteen interruptive forces that most disrupt our thinking.[4] These forces leave leaders, their teams, and their organizations endangered, exposed, and ineffective the more AI improves. Among the fourteen forces, the three forces of persuasion, homogeneity, and polarization are those that most sever our thinking in the Age of AI. In what follows, we will consider each of these three forces in turn.

PERSUASION

The first AI-related interruptive thinking disrupter is *persuasion.* You might be thinking, what's wrong with persuasion? Aren't critique and debate the lifeblood of a healthy society? They are indeed. But the wrong kind of persuasion is where experts try to force your attention toward partial truth. We all use persuasive thinking suppressants from time to time, but I have observed that many AI arguments that grab people by the throat use exactly these systems, especially in relation to purpose and what it is to be human. As literary critic Alan Jacobs wrote:

> Academia and the other high-ranking professions are good at main-
> taining "ideological discipline" within their ranks, and people who
> do well in the academy tend to have "assignable curiosity," which
> is to say, they are obediently interested in the things they're told to
> be interested in.[5]

Today, data science, AI research, and business leadership are high-ranking professions. Are you assignably curious about AI disaster scenarios? Am I assignably curious about human growth? Persuasion can rip cognitive independence from any of us.

"Black Museum" explored an assumption that a person's value is no more than what is reducible to equations. This assumption is similar to the assumption that "you are just your brain"—or, similarly, "your value lies 'only' in your brain." In the context of advancing technology, these assumptions arise frequently and persuasively. The assumptions matter since, in the words of essayist and novelist Marilynne Robinson, "whoever controls the definition of the mind controls the definition of humankind itself." Let's explore these brain-related assumptions from two angles: neuroscience and philosophy. We will see that while it is not simple to explore the assumptions, threshold leaders will invite openness and independent thinking in this domain.

Former neuroscientist Dr. Sharon Dirckx defines "mind" as "the bearer of the unseen, inner life of a person, in the form of thoughts, feelings, emotions and memories . . . the bearer of consciousness."[6] By contrast, "brain" refers to the physical organ in our skulls that has a mushroomlike consistency and weighs about 1.5 kg (or 2.3 kg if your brain weight matches the heaviest ever recorded, that of a US male who died in 1992). According to this description, brains are physical and minds are not physical.

Neuroscientists have performed experiments that may shed light on this distinction. The interpretation of these experiments is controversial, so I do not attach too much weight to them alone. In his 2006 *Science* paper "Detecting Awareness in the Vegetative State," Professor Adrian Owen and others "confirmed beyond any doubt" that a patient who fulfilled the clinical criteria for a diagnosis of the vegetative state "was consciously aware of herself and her surroundings."[7] This study involved asking a patient and a control group to imagine playing a game of tennis and, later, to imagine visiting all the rooms in their house. Researchers measured participants' neural responses using functional magnetic resonance imaging (fMRI). Owen and his colleagues noted that the patient's "neural responses were indistinguishable from those observed in healthy volunteers" and that her "decision [conveyed by virtue of her neural responses] to cooperate with the authors by imagining particular tasks when asked to do so represents a clear act of intention."[8] Owen later observed that what we are seeing is "intact minds adrift deep within damaged bodies and brains."[9] Therefore, Professor Adrian Owen's experiments on patients in the vegetative state may indicate that the state of your brain and the state of your mind are two different things.

In addition, the assumption that you are just your brain often rests on a philosophical view known as materialism, which is the view that the observable physical world is all that exists.[10] The word *observable* is key. Materialism can also include the belief that science is the only way to truth, a belief known as scientism.[11] A significant number of scientists and philosophers think that scientism omits vital parts of reality. For example, neuroscientist and poet Raymond Tallis wrote that "[t]he assumption that 'if science can't see it, then it is isn't real' has nothing to do with science and everything to do with 'scientism'—belief [faith] in . . . the omni-competence of a sub-set of sciences—the natural, rather than the social, sciences."[12] While not all materialists hold scientistic views, Tallis's point is relevant to all materialists, because materialists, too, have belief or faith in what one cannot observe.

All scientists, whether neuroscientists or computer scientists, have belief or faith. For example, all scientists have faith in the view that we can describe the universe by means of laws, the effects of which we can observe. No experiment has established that the view itself is infallibly correct, yet the view is (quite reasonably) widely and strongly held. As Professor Paul Davies, a physicist at Arizona State University, pointed out, "Science can proceed only if the scientist adopts an essentially theological world-view . . . even the most atheistic scientist accepts as an act of faith . . . a law-like order in nature that is at least in part comprehensible to us."[13] If scientists do assume that science is the only way to truth, they are incorrect.

In this context, it is important to consider what is persuasion and what is established fact. Let's consider an argument made by Professor Frank Wilczek, recipient of the 2004 Nobel Prize in Physics. Wilczek accepts that no current AI research enables AI to get close to two "big advantages" that humans have over AI: connectivity and interactive development.[14] Wilczek then goes on to note that these two advantages appear "transient." Wilczek seems to take quite a leap of faith here. His own summary of his argument is:

- "Human mind emerges from matter;
- "Matter is what physics says it is;
- "*Therefore*, the human mind emerges from physical processes we understand and can reproduce artificially;
- "*Therefore*, natural intelligence is a special case of artificial intelligence."[15]

Wilczek correctly states that his first two points are hypotheses; in other words, not currently evidentially supported. His first two points are beliefs or worldview statements, not testable scientific hypotheses. Scientists quite reasonably interpret science according to their beliefs. It is vital that we see the difference between philosophical statements (such as the first two bullet points above) and scientific ones, as this influences the discussion about technology and humanity via our understanding of mind and brain.

Wilczek's first argument ("the human mind emerges from matter") is a negative hypothesis (being accurately restated as "the human mind emerges from *nothing other than* matter). Proving a negative is notoriously difficult and the onus is on Wilczek, as much as others, to make a compelling case. Wilczek has faith that his hypothesis is correct and has not yet proved it. As Professor Samir Okasha noted, it is apparently impossible to answer philosophical questions through science.[16]

At the threshold, leaders resist their and others' thinking being interrupted by persuasion toward a partial truth. Threshold leaders embrace profound truth; as defined by Danish physicist Niels Bohr: "profound truths [are] recognized by the fact that the opposite is also a profound truth, in contrast to trivialities where opposites are obviously absurd."[17] For now, threshold as profound truth means at least three things:

a. Being at ease with, even seeking out, the sometimes contradictory contribution of different disciplines. Threshold leaders are keen to learn about consciousness and related topics from neuroscientists, philosophers, poets, and others. They foster ease by not rushing to solutions and by cultivating ease with complexity.

b. Inviting others into the conversation, regardless of how they currently relate to belief or faith.

c. Nurturing openness and provisionality as to what latest AI and other scientific experiments show. This provisionality is the heartbeat of science, as later discoveries frequently disprove or reshape earlier "certainties." As we approach a more assimilated AI–human future, threshold leaders remain

at best ambivalent about whether AI could ever bridge some
current big gaps to human performance.

One of the finest ways to cultivate such ease, invitation, and openness
is to nurture independent thinking rather than interruption.

HOMOGENEITY

The second AI-relevant thinking disrupter is *homogeneity*. Homogeneity
tramples the diversity of reality and the reality of diversity. In December
2020, CNN reported that Timnit Gebru, known for her research into bias
and inequality in AI, was leaving Google. Two months later, Google fired
the cohead of their AI ethics unit, Margaret Mitchell. Whatever the full
reasons behind both departures, what concerns me is that the quality and
diversity of thinking in a world-shaping organization will decline, to the
extent that they lose access to different viewpoints.

Gebru had previously collaborated with Joy Buolamwini, the self-
professed "poet of code" and "daughter of art and science," to expose flaws
in facial recognition technology.[18] Buolamwini herself later featured in
an award-winning Netflix documentary, *Coded Bias*, which investigated
algorithmic bias.

This bias and these flaws make some people uncomfortable. Com-
menting on Buolamwini's and Gebru's work and addressing the rest of us,
Professor Ruha Benjamin urges us not to run from discomfort.

> So much of what goes under the umbrella of diversity, equity, and
> inclusion is what sociologists call "happy talk." We want to celebrate
> diversity and think about what it gives us as a company or as an
> organization, but when it feels like that diversity is causing trouble
> or holding things up or making work difficult, all of a sudden it's
> not a welcomed difference anymore . . . diversity is supposed to
> make us uncomfortable with the status quo.[19]

When I encountered Buolamwini's, Gebru's, and Benjamin's work, I
realized that it wasn't just me noticing this connection of homogeneity,
AI, and organizational leadership.

Homogeneity is an interruption because, in actuality, homogeneity is a myth. Those who buy into it and build technology on it block leaders and teams from bringing the reality of their differences to their thinking. Interrupting those who see things differently is unhealthy and demotivating, and, sadly, increasingly common and difficult to resist in our technologies. The result is that people are restricted from bringing their full selves to their thinking.

It is worrisome that homogeneity expands. During the last few years, several AI systems characterized women or blacks in sexist or racist ways.[20] These systems drew on data sets and neural networks that were typically uniformly biased. For example, in one data set, women were 33 percent more likely to appear in photographs related to cooking. The problem is not just that machine learning systems mirror such homogeneous biases, but that they also amplify them.[21] The AI neural network trained on these cooking images predicted that the person cooking was 68 percent more likely to be a woman, not just 33 percent more likely, "even when an image was clearly of a balding man in a kitchen."[22] This example resonates with me, as I am a bald man sometimes found cooking Indian curries or Greek roasted vegetables in a kitchen.

According to Benjamin, such biases should not surprise us. In her book *Race After Technology,* Benjamin looked at the historical dimension of race in technology and noted that "design is sometimes discriminatory," and that "knowingly or unknowingly, we embed our human assumptions in tech development."[23] If I assume that you are the same as me or even represent you as the same as me by instructing a data set that you are like me, I interrupt your ability to contribute as your full self. The work of bringing such assumptions to light carries uncertainty and risk, and it may feel contradictory. It is sad that some companies developing AI don't seem to care much about eliminating homogeneity.[24] All the more reason to cross over to threshold leadership.

Crossing over involves contrasting the liminality of the strong with the liminality of the weak. If you are strong, your liminality may be to access humility or passivity. If you are weak, your liminality may be to access assertiveness and resistance. In one sense, this is a simplistic division, as someone who thinks he or she is weak may in fact be strongly oppressing someone else. Also, we all need both humility and assertiveness, not just one of these qualities. But what matters most is to feel and act on the discomfort that homogeneity prompts.

Homogeneity can also be problematic philosophically, in the case where we assume that human and machine purpose are closely comparable. This assumption interrupts our access to our highest purpose, through conflating human and machine aims.

Is machine purpose closely comparable to human purpose? This topic is important because purpose drives high-performing individuals, teams, and organizations.[25] In the context of AI, some seek to equate human purpose with algorithmic machine purpose.[26] But on deeper analysis, one can only equate human purpose with machine or algorithmic purpose by defining purpose narrowly.

To understand this, let's consider what purpose is. John Lennox, professor of mathematics at the University of Oxford, offers this illustration:

> Suppose you ask: Why is this water boiling? I may say that heat energy from the gas flame is being conducted through the copper base of the kettle and is agitating the molecules of the water to such an extent that the water is boiling. Or I may say that the water is boiling because I would like a cup of tea. We see at once that both of these explanations are equally rational—they each make perfect sense—but they are very different. The first is scientific and the second is personal, involving my intentions, will and desire. What is also obvious is that the two explanations do not conflict or even compete. They complement each other.[27]

The first description above refers to mechanistic purpose, which is how something occurs, in the sense of what happened scientifically. The second refers to intentional purpose, which explains why something occurs, in the sense of a desired goal or future state. Here, as often elsewhere, a scientific explanation functions together with, and even enhances the need for, other explanations. Some AI theorists and practitioners collapse their usage of the word *purpose* into the first sense or, worse, blur the distinctions between the two senses.[28]

POLARIZATION

More briefly, we turn to the third thinking disrupter, *polarization*. Polarization is not the same thing as disagreement. Polarization is what happens when human beings detach, disconnect, and disengage from one another, clinging to and entrenching assumptions such as "my values are superior to yours," and "who I am is immutably entrenched in these values." What is at work is an assumption of "core difference," which Kline views as "nothing less than the fear of ceasing to be" and which "turns us into idiots, wild, unable to create the conditions for new thinking between us because we are not interested at all in where the other will go in their thinking. We do not care."[29]

Unless we lead well, AI will amplify the horror Kline describes above. AI-fueled polarization stalks us now more than ever, as seen in state-sponsored bot farms spewing out divisive political memes and in the rise of deepfakes, which make it easy to doctor videos and images. Samuel Woolley, author of *The Reality Game: How the Next Wave of Technology Will Break the Truth*,[30] observed that in the 2016 US general election campaign, "The goal was to divide and conquer as much as it was to dupe and convince."[31] Duping sounds to me a lot like persuasion, which we have explored. In relation to polarization, many fear that Silicon Valley has already lost the battle against divisive social media tactics.[32]

Increasingly fueled by algorithms, polarization interrupts our relationships, as we then ask: Why *listen* to you if I know, or fear, that at some level I might cease to *be* if you say something reasonable? Instead, threshold leaders are inclusive, generous, and generative.

In theory, we could program AI never to persuade, homogenize, or polarize us, but I see no signs of this happening. These three disrupters accelerate the slide of our collective intelligence, as we fail to create environments in which leaders and teams can think effectively. Individuals feel numb, teams stutter, and organizations wither. How tragic that, encountering the great opportunities of shaping a tremendous human-AI world, so many of us trash our equipment. What, then, are the leadership solutions to these trends? At the heart of the answer is the quality of attention you provide to others. It is to this beautiful topic that we now turn.

THRILLING POTENTIAL OF THE HUMAN MIND

Meet one of my clients, whom I'll call Beatrice. She is a US-based senior partner in a major international law firm whose technology practice was considering acquiring an AI start-up. Beatrice had repeatedly earned promotions in an almost 100 percent male context, but came to me saying, "I don't feel like a leader." She told me about how her market was being cannibalized and how she lacked confidence in front of her colleagues and clients. "I feel so insecure in partner meetings when they show off their logical skills. I feel like I disappear," she said, adding, "What must my boss think of me?" Beatrice wasn't sleeping, regularly held more than six hours of back-to-back meetings, and felt anxious about the effects of an accident she had suffered earlier in life. She added, "When I get afraid, I can't think, I can't breathe properly, it blocks my mind."

Eventually, Beatrice found assurance and confidence by using independent thinking and rhythms of rest and performance. What was at the root of this and how did she break through?

Beatrice's inertia and fear arose from not thinking *for* herself *as* herself. She often tried to second-guess what her boss might want, which resulted in derivative thinking, not her own finest thinking. So, I began our work together by assuming that Beatrice could do her own thinking better than I or anyone could do her thinking. When she told me about her fears, I avoided leading her down a line of "progress" that might have worked if her situation were mine (which it could never be). Instead, I felt sure she could make exceptional progress herself. I settled back, giving her my undivided attention, and asked, "What more do you think, or feel, or want to say?"

When Beatrice was stuck, I used her own words to ask, "If you knew you were a leader, how would you contribute?" At the center of my ambition for her was my trust in her own uniquely human intelligence. For years, she had struggled to feel intelligent, despite her success. I think that she not only heard this ambition and trust from me but felt it, too. Although sometimes I said little, the quality of how I treated Beatrice gradually helped her bring her full self, not just her fearful or logical self, to her thinking in her daily work.

Beatrice's breakthroughs made a world of difference to her work outcomes. Her team's engagement scores and the company's market share grew. Senior colleagues reported that Beatrice shone in meetings. She received her first invitation to represent her company at an industry conference. More than this, she felt satisfied and said to me: "I feel like more of a leader. I'm realizing that my role is an amazing platform that I can use even more." As AI was encroaching on her market, Beatrice was becoming more relevant, influential, and loved. She was stepping onto a threshold. This led to her quoting words written by Scott Russell Sanders when describing her feelings about her transformation: "Joy banged in my ribs!"

Whatever leaders think they are doing when they lead, they are revealing something of their being. Previously, Beatrice revealed her anxious self as she led, suffocated by doubt. After our work together, she revealed her purposeful and centered self as she led, journeying to "family" leadership and even a little beyond. In her new story, Beatrice knows that she is not like AI, a disconnected rational algorithm, but rather a present, loving, generative, independent thinker. Having embraced the challenge of connecting her thinking with her being, she is now helping her teams, organization, and industry do the same. What the human mind can do on its own never ceases to amaze me.

A GENERATIVE PIONEER

The path of nurturing independent thinking is deeply rooted in the Thinking Environment®, which was pioneered by author and coach Nancy Kline. In 2015, I first read Kline's book *Time to Think*. Rarely has a book so upended my thinking. It made it to my "top ten books that everyone should read" list, along with George Eliot's *Middlemarch* and Vikram Seth's *A Suitable Boy*. Kline's book explores the soaring potential of human thinking. When I read it, I was immediately hooked and dabbled with some of her findings as I coached leaders. The impact of my adjusted approach impressed me, so I signed up for Kline's coach training program, waiting a full year for a spot.

I still remember when I first met Kline in 2017. I had journeyed by train from Oxford to the picturesque village of Goring-on-Thames, arriving shortly after breakfast at Friars Ford, the location for our training. Friars Ford is a Victorian Gothic former family home. As I stepped into the grand hallway, I already felt inspired by my journey through the countryside, inspired by my walk up the bluebell-lined driveway, and inspired by the elegant interior decor. Kline emerged from the drawing room. Poised, elegant, and smiling, she approached me.

Have you ever met someone whose work has deeply influenced you? There's that moment of awkwardness before either of you speaks. There are those two seconds (or was it two days?) of wondering what opening profundity I could offer up that could possibly match the depth of what I had read in *Time to Think*. Well, the awkwardness and wondering were mine, not hers. From the moment she greeted me with a warm New Mexican "Hello" and a hug, and then for the next three days we spent together, I felt even more inspired.

Why is Kline special as a coach and leader? Slight in build, she is one of the most powerful people I know. She gives a greater quality of attention than I have experienced anywhere else. This attention is a multifaceted thing, and Kline defines it as follows: "Listening without interruption and with interest in where the thinker will go next in their thinking."[33] In her case, her attention glimmers, sparkles, and invites worlds of thinking to unfurl. This quality of attention and these abilities are characteristic of threshold leaders.

In the last five years, I have become more and more deeply impressed by the potential of independent thinking to change the world, especially as technology advances. My colleagues and I regularly share our passion for this way of being with business leaders at pharmaceutical, technological, industrial, and other companies, who incorporate the practices. The more we reflect on leadership in the Age of AI, the more we are realizing that independent thinking is one of the core pathways that will help human leaders succeed. Why is this?

Core to Kline's work is the observation that the quality of your independent thinking drives the quality of your decisions and actions. This is valuable work. According to Kline, thanks to Thinking Environment* the president of Zambia saved 40 percent of the GNP (gross national product) in thirty-six minutes, a company rescued a $200 million product in forty-five minutes, and a hospital moved in nine months from one star to four.[34]

In Kline's own words, "[A leader's] first job, their forever most important job, the job that they have, is to create the conditions for stunning independent thinking in every life they influence."[35] At the threshold you are an independent thinker, defeating interruptive technological forces. You create bountiful attentional space for yourself and others, and you enjoy something that may sound contradictory: integrated cognitive autonomy. I wonder how much more effective our AI initiatives could already have been if engineers, entrepreneurs, educators, policymakers, and investors had encouraged truly independent thinking. We don't have to intend AI or our own processes to interrupt us.

At the core of the Thinking Environment˙ is an assumption that we do our best thinking when we bring our whole being to it, rather than subverting our thinking to someone else's being, ignoring emotion as a component of thinking, or neglecting physical factors in thinking. In these observations, we can see that the various parts of this book interrelate.

It is sad that most of us have been trained not to think for ourselves, an irony where we view ourselves as pioneers and thought leaders. When we first learned to speak, our parents finished our sentences for us, assuming they knew what we were about to say. When we debated in class at school, our peers (and too often our teachers) interrupted us. When we started paid work, we learned to think what our superiors wanted us to think in order to get that bonus, that promotion, or that other badge of attainment. By contrast, the human mind can achieve great things on its own, when in a conducive environment.

THRESHOLD RESOURCES

RESOURCES FOR ANY LEADER

Resource 6. Pre-mortem Bias Reducer
PURPOSE
- Invite everyone's thinking in meetings.
- Challenge persuasion, homogeneity, and polarization.
- Assess and escape subservient or derivative thinking.

PROCESS
Fifteen-minute survey and discussion resource.

The quality of thinking in your organization can be one of your most important assets, yet bias can hamper results in many ways, not least via thinking disrupters such as persuasion, homogeneity, and polarization. As we increasingly rely on data sets and neural networks that form part of machine learning systems, the risk of such biases will increase.

As you use this resource, not only will meetings go better, it will change who you are as you attend meetings. In other words, this resource can help shift team culture toward inviting everyone's best thinking in meetings. This will in turn minimize the chances of a range of these biases taking hold.

This resource consists of two parts:
  I.   A MEETING EVALUATION SURVEY
  II. A WAY TO PROCESS THE SURVEY RESULTS

I. A MEETING EVALUATION SURVEY
    Have each team member score the following questions in relation to a recent meeting. Answer the following questions using a scale of 1 to 7 (1 being "not at all" and 7 being "completely"):
    - To what extent did we need this meeting?
    - To what extent did we separate administrative, tactical, strategic, and developmental discussions appropriately?
    - To what extent were updates focused on what was critically important?
    - To what extent were we free from hurry?
    - To what extent did we avoid obedient thinking in this meeting?

- How much do I think the chair cherished my independence of thought in this meeting?
- To what extent did this meeting include a diversely comprised team?
- How effectively did our discussions draw on various kinds of diversity that exist among us?
- To what extent did I have as much airtime as others?
- To what extent did I feel safe enough to voice what I really thought about every topic addressed during this meeting?
- How much did we really listen to one another during this meeting?

Also, provide a free-form box in the survey for respondents to enter qualitative comments if they wish.

II. A WAY TO PROCESS THE RESULTS

Arrange dedicated time for your team to discuss the results. You might do this by following a five-step process:

1. Have each participant answer the survey well in advance.
2. Collate the results in an accessible format. The survey questions are grouped as follows:
    a. Questions 1–4 relate to general meeting effectiveness.
    b. Questions 5–6 relate to persuasion.
    c. Questions 7–9 relate to homogenization.
    d. Questions 10–11 relate to polarization.
3. Share the results with team members in good time ahead of your next meeting.
4. Schedule time during a forthcoming meeting, in which each person has time to say what they think about the results. During this discussion, widen the aperture to include your organization by asking questions such as:
    a. How effective is our meeting culture?
    b. How masterful are we at removing thinking disrupters, especially interruption?
5. Decide together what you want to do about it. For example, what processes will you keep or change?

PAYOFF
- Rectify failures before they happen.
- Generate better AI ideas. Implement them better.

CHAPTER TAKEAWAYS

- The mind is an exquisite instrument.
- Homogeneity, polarization, and persuasion disrupt the quality of leadership thinking.
- Connect your thinking and being using independent thinking, thereby defying AI's imminent choke hold on your thinking.
- Raise the intelligence of groups by paying magnificent attention, letting go of interruption, and cherishing difference.

5

AN AUTUMNAL
TUESDAY IN BERLIN

Understanding evolves through three phases:
simplistic, complex, and profoundly simple.

—WILLIAM SCHUTZ

Independent thinking is a way of being that I find peerless and profound
in a digital environment. Building on the work of Will Schutz and others,
organizational theorist Karl Weick described three stages that people experience when they make sense of overwhelming events or ideas: superficial
simplicity, confused complexity, and profound simplicity:

> On the far side of complexity lies profound simplicity. These sim-
> plicities may sound a lot like the near-side superficial simplicities
> that you and others started with. But that apparent similarity is
> deceiving. Profound simplicities mean something very different.
> They are seasoned simplicities, simplicities that have been tested by
> mentally simulating their consequences, simplicities that reaffirm
> what it means to be a human being.[1]

What does independent thinking look like in practice? In the following annotated story, we will see how a Thinking Environment* transformed the way I felt and performed at a moment of loss. The numbers
in the story will be important later.

It's an autumnal Tuesday evening in Berlin in the year 2018. I arrive at a hotel to take over from Brian Draper, my friend and colleague, who has set up a labyrinth on behalf of a new client.[2] I feel tired from a long day and the journey, but I'm exhilarated to soon shepherd participants into the soulful space that Brian's curated. I bound up the glass staircase and put my rucksack on one side, as a participant wanders up, poised to enter the ballroom containing the labyrinth. As is our practice, I guide the participant inside to the reflective space. I enjoy the music and then return to the hall outside. As I start to settle in for a long night of helping people into the labyrinth, I slip my hand into the main pocket of my rucksack to pull out my laptop. But it isn't there.

Strange, I think. I must have got it out a few minutes ago and forgotten or left it in the taxi, plane, or airport somewhere on my trip from London. But I just finished this journey and knew I hadn't used my laptop at all. It should still have been in my rucksack.

Slowly, crushingly, I realize that someone had stolen my laptop while my back was briefly turned just a few minutes earlier. During the next half hour, several of us learn that two other laptops had been stolen at the same time, one along with a handbag, which also contained a colleague's passport with visas, a wallet, an iPad, and an iPhone. Two thieves had loitered in the hotel, waiting for an opportunity.

I felt frustrated, shocked, sad, and angry. I also had a job to do. Participants were arriving to walk into a labyrinth experience that they suspected (rightly in several cases) would be life-changing. How could I center myself?

Answer: I didn't, not immediately at least. I had two challenges: focus quickly to be able to serve those yet to experience the labyrinth, and also come to terms with the imminent days-long administrative mountain that accompanies getting a new laptop. This latter challenge decentered me for days. My turning point came the following Tuesday when my friend Georgie Lyttelton coached me.

In her life and work, Georgie commits to helping others think for themselves [1]. Sipping a glass of cool water in that session [2], my thoughts swirled around for nearly an hour before finally starting to settle. I felt free and relieved as I spoke, paused, spoke again, paused for a long time, and only sometimes (in this process) asked for Georgie's help to keep my thinking going [3].

The quality of attention that Georgie showed me transformed me in this moment. Wherever my eyes went, hers were on mine. Whenever I stopped talking, she didn't assume that I had stopped thinking. Viscerally, perhaps subconsciously, I knew she was there for me, rooting for me to do my best thinking [4].

Until this conversation, I hadn't realized how stressed I was about completing my accumulated workload. Reflective thinking had eluded me for a whole week. I now saw that I'd been glazed and distracted in several recent interactions. I was about to go into a thirty-six-hour period with nearly ten separate meetings or calls. Now I felt sad, angry, and regretful. Georgie knew that these feelings did not inhibit my thinking—far from it [5]. As a result, I felt secure to continue to think for myself, rather than feeling unsafe [6].

Georgie asked me questions such as: What more do you think, feel, or want to say? If you knew you were an artesian spring for others, how would you be in the next thirty-six hours? [7] From the latter question, I generated an insight that liberated me. Finally, although I was doing the vast majority of talking, Georgie noted that we are equal as thinkers and that the differences between us add quality to our thinking in the session [8]. We ended the conversation, as we had so many others, by appreciating a quality about each other [9].

This conversation enabled me to be there for each person in my upcoming meetings and calls, fully present, fully and lovingly attentive. Frustration turned to joy, anger to flow, distraction to focus and creativity. Now I felt delighted to be ordering my new laptop and pressing into my future.

How did Georgie create such an environment for me? Why does it matter in an Age of AI? The following numbered sections correspond to the numbers above and address these two questions. The numbered sections also benefit from a rich warehouse of knowledge developed by Nancy Kline and her colleagues, drawing from deep reflection on decades of observing what happens when people generate waves of thinking and what happens when they pause. They have codified this warehouse into learnings that, from research and experience, are highly effective for individuals, teams, and organizations.[3] Below, I lay out some of these findings, grouped in ways that will be most significant for leaders in the Age of AI.

[1] THREE PROMISES WE MADE

Truly independent thinking requires three promises. The thinker (me, in this case) promises to think for oneself. The thinking partner (Georgie, in this case) promises, first, not to interrupt me and, second, to be more fascinated by what I say next than she is by how she could respond. These are acts of loving attention to self and other.

A key part of being fascinated by what someone else will say next is to trust that person's intelligence as a thinker. This trust accords with Carol Dweck's groundbreaking research at Stanford University, in which she found that "when children were recognized for their efforts to think, they created a belief, and then a reality, that intelligence grows."[4]

This fascination and these promises are so resolutely *un*-digital, so *un*-mechanical, so *un*-subverted. In the Age of AI, leaders will do their best independent thinking when they bring their whole being to it: brains, mind, body, and soul.

Maybe you're thinking, *No one can truly think independently, because we are all connected.* I agree that we are all connected. For this reason I view independent thinking as akin to what some philosophers call *autonomous thinking*. Autonomous thinking includes the idea that healthy, belief-forming environments are those in which beliefs are formed, at least in part, as groups.[5]

In this section, as well as in all the following numbered sections, I will encourage you to think independently about the threshold by posing at least one question.

If you knew that your independent thinking matters, what would change?

[2] PLACE

Georgie cared that I was physically in a place that communicated to me that I matter. "Place" referred to my physical location, mind, and body. I chose to declutter or refresh all three by clearing my desk, shutting off my email, and getting a good night's sleep before our conversation (as well as by pouring myself a glass of water during the conversation itself).

This reminds me of the Japanese minimalist concept of *ma*, which honors the space between things. "That space can often be crowded out by clutter," wrote Brian Draper. "Clutter fills a void, carelessly, and can stifle not just our physical space, but with it our peace of mind, and our room to think and thrive, too."[6]

Many AI writers focus understandably on how far AI can go beyond its already impressive achievements. My fascination is with how far humans can go beyond our already impressive achievements. Threshold leaders understand that place lies at the heart of such human progress. In the coming years, AI-fueled digital implants may supercharge our human abilities and/or may imprison us more than our smartphones already do. In either case, cultivating inspiring places in yourself and externally will remain a key differentiator.

If you knew that the state of your mind and body regulates your outcomes, what would change?

[3] WAVES AND PAUSES

Georgie understood that thinking seems to come in waves and pauses. Often, a pause is a midwife to further waves of thinking, rather than an invitation to the listener to provide a "magic bullet" answer, as if that could solve everything. Often, when I paused, the last thing I needed was Georgie chiming in with an assertion or a question, even if she did me the honor of not interrupting. At other times, my pauses were full stops, moments where I needed help to continue my thinking. In line with our agreement, I asked verbally for help at these times, and Georgie offered a question that enabled me to resume thinking for myself.

The end-of-chapter resource "Thinking Pairs" provides a human template for what this looks like in practice. Thinking in waves and pauses is one of the most exhilarating recent findings about independent thinking.[7]

If you knew that you listen to yourself in your pauses, what would change?

[4] CULTIVATING A RAW IMMEDIACY OF PRESENCE

Georgie listened with great respect. In our conversation, she did not interrupt and cultivated a "raw immediacy of presence," in Cynthia Bourgeault's words. In the face of challenges of climate change, polarization, and pandemic, AI could be a miracle that humanity needs. But to access this miracle, we need to face up to the deathly nature of the thing AI and humans both host: interruption. Let's reflect more on interruption, as this is central to the quality of presence I'm talking about.

I find interruption useful or beautiful in only two, possibly three, situations. First, in an emergency, for example, if the building is on fire at this very moment. Second, in the interrupting cow joke.[8] Third, some coaches regularly use interruption as a deliberate coaching method to try to break what they see as unhelpful patterns in a client or to blurt something into the conversation. I have not had the opportunity to assess the efficacy of this method, for which some claim a basis in neuro-linguistic programming.

Interruption costs. In her book *What Happened,* Hillary Rodham Clinton shared an arresting example:

> Arianna Huffington was recently interrupted in a meeting of the Uber board of directors when she was making a point about—of all things—how important it was to increase the number of women on the board! And the man who talked over her did so to say that increasing women would only mean more talking! You can't make this up.[9]

As the playwright Albert Guinon said, "There are people who, instead of listening to what's being said to them, are already listening to what they're going to say themselves."[10] Instead of reflexively interrupting people, I invite you to take time to carry out a risk assessment. Consider, on the one hand, what you gain by interrupting people—for example, you insert into the conversation your point of view, much of which you knew before you said it. On the other hand, you know extremely little of what you'll lose when you interrupt, for the simple reason that you don't know what others were going to say next. Even if, as sometimes happens, you successfully guessed the rest of their sentence, you still didn't know where they could have developed their thinking from there. How many people would sign off on an investment for which they know more about the benefits than the costs? In our conversations and digital lives, we often sign off on extremely unwise risk assessments.

Purely at the level of not interrupting, AI could do well. Digital assistants already exist that do not interrupt and only offer questions or input when invited. And AI could also pose some good questions.

If we have learned one thing about the development of technology over recent decades, it is that we tend to institutionalize systems of interruption. In the future, some leaders may develop AIs with even more interruption in mind, as shortsighted business models demand it. For example, imagine an AI that interrupts your thinking to offer advice or ask a question, based on powerful predictive engines. You may have already noticed this happening via your devices. The advice may be well directed, but in offering it the AI just cut across whatever thoughts you were about to have. If attentiveness is the natural prayer of the soul, as fifteenth-century French philosopher and theologian Nicolas Malebranche put it, then some algorithms threaten to hijack our prayers.

But just because we have often developed interruptive technology doesn't mean we must continue to lead it this way. At our best, human purpose involves a wild fascination with what others will say next. Threshold leaders create the conditions for such fascination, including technologically.

Georgie possessed a raw immediacy of presence. This immediacy was not purely due to her not interrupting. Humans seem uniquely able to show a sublime quality of presence. It was this that helped me break through.

Leaders have few better resources than this attentive presence. As Dr. Otto Scharmer (more about him in chapter 9) explains in his work on transformation, "All real creativity, all profound innovation, and all deep civilizational renewal are based on the same source: the capacity for sustained attention." In such profoundly attentive settings, the thinker senses the following: "My whole being has slowed down. I feel more quiet and present and more like my authentic self. I am connected to something larger than myself."[11]

I love how psychiatrist Ian McGilchrist puts it: "Attention changes the world. How you attend to it changes what it is you find there."[12] This was exactly the sort of change and interior condition I experienced in my post-theft conversation with Georgie. In a world of increasingly advanced machines, the ability to create such conditions will set humans apart.[13] Done well, attending draws on your whole being in a way that will sustain and even increase your relevance in the Age of AI.

This quality of attention, even loving presence, will be essential if we want to survive and thrive in a world of AI. Maybe you prefer an adjacent term such as *positivity resonance* or *unconditional positive regard,* the latter term most associated with psychologist Carl Rogers. However you frame it, an exceptional quality of attention will be increasingly critical, and I cannot see AI gaining it any time soon. Aspire to this quality of attention in your interactions with others.

> *If you knew that your team is fascinated by what you will say next, what would change?*

[5] FEELINGS

Leaders who allow appropriate emotional release in others will help those people, and the ones around them.[14] Georgie did this by not being judgmental about my anger, sadness, and regret. Alan Jacobs explains something that the neuroscientist Antonio Damasio discovered:

> When people have limited or nonexistent emotional responses to situations, whether through injury or congenital defect, their decision making is seriously compromised. They use reason alone—and, it turns out, reason alone is an insufficient guide to action. . . . Learning to feel as we should is enormously helpful for learning to think as we should.[15]

Chapter 1 addressed the difficulty of an AI "having" emotions in the same way that humans "have" them. A related question is, Could an AI "welcome" another human's emotional release in the same way that a human "welcomes" it? This is a complex question. Some might argue that empathy and sympathy are lesser where they are programmed. When Siri tells me, "I'm sorry you feel that way," I don't sense that Siri empathizes. This is a worry I have when it comes to machine-created empathy. They're faking it. Even though the apology "I'm sorry you feel that way" is sometimes viewed as the most infuriating apology ever, it is possible for a human to offer it in a genuine, inviting way.

The loving ambition and loving attention that Beatrice felt from me and that I felt from Georgie goes beyond what advanced AI could provide. Even futuristic algorithms could "press the button" and generate some quite effective questions and other statements, but would lack the human understanding and connection that seem to be critical to help others generate their own whole thinking. Research is needed to validate or disprove this, but my assertion here leans partly on the findings about multiple intelligences presented in chapter 1. No theoretical basis exists for AI to go beyond general cognitive intelligence and match or surpass humans in emotional intelligence or wisdom. So, AI may never get close to humans in generating the connection that often underpins breakthrough dialogue.

Feelings are a core part of being. This is an example of the kind of integrated thinking we need if we're to shape a magnificent future. The path of thinking independently isn't just about unleashing brilliant ideas. The path emerges as a way of also helping the mind to settle, with all the attendant, holistic, transformative benefits of that.

If you knew that others welcome your feelings, what would change?

[6] BEING AT EASE

During our conversation, Georgie didn't hurry to get anywhere else. She knew that "ease creates; urgency destroys."[16]

In some ways, increasingly advanced AI could be more free from internal urgency than humans. Fear of being usurped by competitors can drive humans to get things done more quickly than required. If AI does not suffer from this kind of temptation, it won't rush others in related thinking processes and could therefore create more ease in some cases. However, the beautifully un-digital, generative centeredness that humans can show when they lack internal urgency still stands apart.

I have a problem with the idea of doing "philosophy with a deadline," in Swedish-born philosopher Nick Bostrom's famous AI-related dictum. At one level, it preserves an important and urgent truth: Given the pace of technological development, it's possible that if humans don't align sufficiently on certain major philosophical, ethical, and other questions, then increasingly nonhuman intelligences may remove much of our agency from us. The problem isn't in longer-form explanation, but in the short-form, four-word dictum. Overly tight deadlines promote rushing and therefore loss of thinking quality. Ease is the opposite of this, as Georgie's ease with me created the conditions for my breakthrough. Having limited time is okay, but unless the building is on fire, let's not rush. Nowhere is ease more important than in humanity's charting of AI's profound ethical and philosophical waters.

Georgie's lack of urgency also helped me commit to action and own that commitment. Personal ownership is, of course, critical to effective leadership.[17] The neuroscientist Professor Paul Brown argues that Thinking Environments* represent

> A special case of establishing limbic resonance . . . the Thinking Environment is a method not just for calming the amygdala, but also for accessing the mind. What happens is that the Thinking Session process creates relationship, mobilises the brain's energy and makes it possible for the person to access all kinds of information that have been below the threshold of working consciousness.[18]

Georgie's ease helped me experience this.

If you knew that you are not in a hurry, what would change?

[7] GENERATIVE QUESTIONS

Georgie asked me incisive questions such as, "If you knew you were an artesian spring for others, how would you be in the next thirty-six hours?" and other generative questions such as, "What more do you think, or feel, or want to say?" Incisive questions free "the human mind of an untrue assumption lived as true."[19] In asking me such questions, Georgie created a breathtakingly strong foundation for me to advance my thinking. Where I had assumed that life was now less enjoyable and that I would have less impact simply because my laptop had been stolen, Georgie helped me face unpleasant trade-offs and focus on what really matters.[20]

I am reminded of the apocryphal story about a consultant who was called in to help fix a nuclear reactor that was in danger of going into meltdown. The consultant took a few minutes to look around the control center, pressed one button, and declared the problem fixed. Weeks later, the company received the consultant's invoice for $1,000,001. Aghast at the high amount for just a few minutes' work, the company asked the consultant to justify his rates. "The fee is one dollar for pressing the button," the consultant explained, "and $1 million for knowing which button to press." The more you pose generative questions, the more you are pressing the right button.

Generative questions are a vital leadership resource, as they are the best way I know to identify and remove assumptions that limit cross-functional teams working together. These questions differ profoundly from assertive inquiry, as often used in leadership development.[21]

Even if clumsy at first, AI systems could in principle ask questions in the "What more?" category. AIs will probably increasingly factor in verbal and visual feedback to improve how effectively they ask these questions. We could also program machines to pose some Thinking Environment[*] questions specified by a logic tree. But threshold leaders use logic as well as judgment and intuition to discern what to do and say in a pause. In this context, logic trees will disappoint.

All the closing questions in sections 1–9 of this annotated story are generative. Some additional generative questions follow:

If you could trust that your children will be fine, what would you do with the rest of your life?

What will you learn about yourself at your next review that you know already? (That's right, what will you learn about yourself that you know already? If something in this question appeals to you, I encourage you to sit with it and see what arises from your subconscious.)

If you knew that only you could define it, what would a successful AI future be?

If you knew that Superintelligence emerges within ten years, what would you do now?

What can you uniquely do that our AI world of tomorrow needs?

Where does your deep gladness meet the world's deep needs?

If you knew that they won't reject your ideas, what would change?

[8] A BINARY STAR: EQUALITY AND DIFFERENCE

In our conversation, Georgie championed equality and difference. Equality means that everyone matters and that everyone's thinking is valuable. Difference refers to the fact that diversities between us add quality to our thinking together.

I view equality and difference as a binary star, as gravitationally joined components. A binary star is comprised of two stars, with neither hosting the center of gravity of the whole binary star. Effective, inspiring leaders require both stars. Equality without difference can lead to an artificial smoothing out of unique and delightful differences among people. Difference without equality can lead to privileging of some people above others. Neither is individualism conducive to good thinking. Homogeneity is a key trend standing in the way of the binary star.

The binary star facilitates breakthroughs, as it hints at justice. As Radecki and Hull noted, drawing on the brain studies published by Cheng and others in 2017, "When we perceive something as 'unfair,' an area of the brain called the insula is activated . . . [which] deals with the extremely important primary emotion of disgust, which compels us to be repulsed."[22] Perceived unfairness repulses most people.

Turning to difference, the brain often treats strangers as a threat, "categorizing them as foes" and processing them as part of an excluded group.

Paying attention to the binary star helps us override our built-in subconscious "out-group bias" by "finding commonalities with them."[23]

The binary star is a beautiful part of the threshold. The binary star shines a light on the fact that when you are what makes you different, you contribute in the grand scheme of things to making me who I am, because I am only what I am in contrast to others. In short, threshold leaders know that we are all connected, as one tapestry.

It may turn out that AIs complement humans in nurturing parts of the binary star. For example, what if AI helped us solve the diversity disaster in the AI realm, in which way too few AI research staff and tech company employees are female or black? Of course, we can question whether humans will "matter" to AI in the same way that humans "matter" to humans. But human history is a litany of systematically sponsored injustices. Whatever direction we pursue, I am convinced of this: The most effective human leaders will cherish the binary star in how they collaborate with and create AI. To start on this journey, you might use the "Binary Star" resource at the end of this chapter.

If you knew that no one can do your thinking
as well as you can do your thinking,
what would change?

[9] APPRECIATION

There's an old rhyming couplet from an anonymous author that goes, "Once I did bad, and that I heard ever; twice I did good, but that I heard never." By contrast, during our coaching conversations, Georgie and I practiced regular mutual appreciation.

Appreciation is also not mere flattery. As Dale Carnegie described these two terms: "One is sincere, the other insincere. One comes from the heart out; the other from the teeth out." On this occasion in October 2018, appreciation created a sincere foundation where I felt willing and able to bring my full being to the conversation, which enabled steps forward in my thinking.

Appreciation's source is the heart. For this reason, I think it will be one of the last human capabilities that AI will match, if it will ever match it.

Many research studies have investigated the emotional impact on recipients of varying mixes of appreciation and critique. The question being asked in these studies is: Other things being equal, in order to maintain a neutral emotional mood in another person, what ratio of appreciation to critique is required? Results vary in these studies, from 5:1 up to as high as 14:1.[24] In other words, if you want to stand a chance of maintaining a neutral emotional mood in your organization, everyone needs to appreciate others at least five times as often as they critique. In how many organizational cultures have you encountered this ratio?

Here are some ideas to help you improve your ability to appreciate:

- Find at least one thing to appreciate about work colleagues each day, and tell them.
- When you spend time with others, appreciate the time, effort, and heart they invest *much* more often than you criticize.
- Appreciate qualities in other people that relate to their *being*, not just things they did (which relate to their *doing*).
- Practice Nancy Kline's three S's of appreciation: succinct, sincere, and specific.
- Write this out and stick it to your fridge or wall: "Silent gratitude isn't very much use to anyone" (author and literary critic Gladys Bronwyn Stern).

If you knew that appreciation improves performance, what would change?

A beautiful feature of independent thinking is its universality. I have seen hundreds of people from many walks of life soar in independent thinking. Although it takes will and practice, like Beatrice or me, you can thrive as a threshold leader using this pathway if you choose.

Independent thinking is an evolutionary leadership practice for individuals that will help you lead the way toward a new dawn at such a difficult societal and political time as this. The opportunity for threshold leaders is also collective. In the next chapter, we explore independent team and organizational thinking.

THRESHOLD RESOURCES

RESOURCES FOR ANY LEADER

Resource 7. Thinking Pairs

Thinking Pairs were pioneered by Nancy Kline (Time To Think, Ltd.). I developed this version of the resource in conjunction with, and by permission of, Kline. For more information, see Nancy Kline's 2020 book, *The Promise That Changes Everything: I Won't Interrupt You.*

PURPOSE
- Generate fresh creative ideas.
- Dissipate confusion.

PROCESS
A thirty-minute paired exercise.

> To be interrupted is not good.
> To get lucky and not be interrupted is better.
> But to know you will not be interrupted allows you truly to think for yourself.
>
> —NANCY KLINE

Thinking Pairs open up a rich universe of possibilities. This resource proceeds from the insight, as described by Kline, that "the quality of your actions and decisions flows from the quality of thinking you do first." I have found this resource immensely valuable and life-giving in my marriage, among friends, and in all manner of business settings. Its power comes from a combination of transformative listening, a profoundly simple setup, and a promise to think for yourself. Enjoy the expertise that you will gain from this exercise!

- Find a partner. Inspire a colleague to join you by asserting that when you listen, your customers and clients think you are brilliant.
- Arrange a time and place to meet, whether physically in the same room or virtually. Do all you can to ensure that

your locations remain as quiet as possible. This includes, for example, switching off digital devices, making arrangements for childcare, and putting a "busy" sign on your office door—whatever it takes to create a good thinking space for you.

- Decide which of you will have the first turn as the Thinker. The other person is the Thinking Partner. After the first Thinking Pair, you will swap roles.
- Set a timer for twelve minutes (the length of each Thinking Pair).
- When the Thinker is ready, indicate this to the Thinking Partner, who then asks the following question: "What do you want to think about and what are your thoughts?"
- The Thinking Partner then does not speak again, at all, for the whole twelve minutes, unless explicitly invited to do so by the Thinker.
- At some point during the twelve minutes, as the Thinker, your waves of thinking may come to a stop. That's okay and perfectly normal. If you need help getting going again with your thinking, ask for help by saying something like "I'm done," or "Over to you," or "Ask me a question." At this point, and *only* at this point, the Thinking Partner should ask this follow-up question: "What more do you think, or feel, or want to say?"
- Ask for this follow-up question as many times as you wish during the twelve minutes. If twelve minutes seems like a long time for you, remember that no one can do your thinking for you. Luxuriate in the time that is given to you. Enjoy breaking free of our usual cultural bonds of superficial or derivative thinking.
- When the timer goes, draw your thinking to a close. (Only the Thinker may end the Thinking Pair early, if he or she wants to.) As the Thinking Partner, maintain confidentiality by not discussing any of what the Thinker said with anyone, ever—even if one-on-one with the Thinker again. The only exception is if the Thinker explicitly raises the topic again and/or gives you permission to discuss it.
- Swap roles. Set the timer for twelve minutes and repeat.
- Close the session by sharing a phrase or sentence of appreciation with each other.

Tips to Use this Resource Well

THINKER	THINKING PARTNER
Be at ease doing your own thinking.	Keep your eyes on the eyes of the Thinker as he or she speaks.
Know that thinking for yourself is still a radical act.	Pay magnificent attention to the Thinker.
Pause. Close your eyes. Look around if you want.	Do not interrupt, for anything (unless the building is on fire).
Ask for a question when you want one.	Only ask follow-up questions when explicitly invited.
Know that you are intelligent, valuable, and lovable.	Be fascinated with what the Thinker will say next.
Know that you can lead and make a difference.	Trust the intelligence of the Thinker.
Know that you can survive and figure this out.	Know that just because the Thinker stops talking doesn't mean he or she has stopped thinking.

Payoff

- Bring your whole being to your thinking.
- Feel more connected to a friend or colleague.

RESOURCE FOR LEADERS IN LARGE ORGANIZATIONS

Resource 8. Binary Star

Purpose

Raise the quality of thinking in your organization by cherishing difference and equality.

Process

A world-café-style, ninety-minute, facilitated module suitable for a large AI-related event. Structure this event around three breakout discussions (table discussions), bookended by plenary opening and closing sessions in which you set context and draw out learnings. In each breakout session, invite a volunteer to share the input (either verbally or written) and then explore the discussion questions together.

Materials required: large venue, tables set out café style for six to eight people, flip charts, preprinted topic sheets, paper tablecloths, pens (or virtual equivalents to all of this).

Refer to the definition and discussion of the "binary star" of difference and equality, earlier in this chapter. This binary star is relevant not just for humans in relation to gender and race, for example, but also for humans as compared with machines. If you view AI as superior, you may impoverish human contributions. If you undervalue the contribution of AI to your work- force, your organization may never get to scale.

TABLE DISCUSSION A
Shining the Binary Star on Our Workforce

- Input: An article in the *London Times* explored the importance of diversity and equality in a world of AI. In April 2019, Simon Duke wrote, "According to a research paper from New York University last week, the 'diversity disaster' in the AI realm risks perpetuating gender and racial biases within society. Women comprise only 15 percent of AI research staff at Facebook and 10 percent at Google, according to the study. For ethnic workers, the picture is much worse; only one in forty Google employees is black."[25]
- Discussion questions:
  - How diverse is our organization today? Consider as many dimensions of diversity as you can think of, as a group.
  - What actions can we take to improve both diversity and equality in our human workforce, not just one or the other?
  - In what ways do humans outperform AI? In what ways does AI outperform humans?
  - If something goes wrong, who should we blame? Only humans? An AI? Why?

TABLE DISCUSSION: B
SHINING THE BINARY STAR ELSEWHERE IN OUR ORGANIZATION
- Input:
  - Many organizations, such as Elon Musk's OpenAI, Nick Bostrom's Future of Humanity Institute, and MIRI work on AI safety and values. The OpenAI website states: "The goal of long-term artificial intelligence (AI) safety is to ensure that advanced AI systems are aligned with human values—that they reliably do things that people want them to do."[26]
  - Alphabet's DeepMind is attempting to build advanced AI with positive human values embedded, so that those values will be retained even long term.[27]
- Discussion questions:
  - How will we agree what positive values we want as an organization?
  - What values are most important in our AI initiatives?

TABLE DISCUSSION: C
PLAYING OUR PART TO SHINE THE BINARY STAR ON WIDER SOCIETY
- Input: Share this commentary from the AI Now institute: "At their best, AI . . . can . . . reduce both conscious and unconscious biases. However, training data, algorithms, and other design choices that shape AI systems may reflect and amplify existing cultural prejudices and inequalities. We already have evidence of these problems. . . . When machine learning is built into complex social systems such as criminal justice, health diagnoses, academic admissions, and hiring and promotion, it may reinforce existing inequalities, regardless of the intentions of the technical developers."[28]
- Discussion questions:
  - What more can we do to understand and mitigate the effects of bias in AI systems?
  - What differences do we smooth over?
  - Who are we treating unequally?
  - What courageous steps could we take to increase fairness around us, for example, by volunteering our AI capabilities?

PAYOFF

Add quality to your discussions, decisions, and actions about AI.

CHAPTER TAKEAWAYS

- Independent thinking is peerless in a digital environment.
- When you nurture independent Thinking Environments®, everything changes. You pave the way for others to bring their whole selves to their thinking, with sometimes surprising, almost always effective, and satisfying results.
- At the threshold, you can encourage independent thinking as follows:
  - Promise to think for yourself, understanding that thinking occurs in waves and pauses.
  - Promise not to anaesthetize dialogue by interrupting. Pay magnificent attention to other people instead.
  - Promise to be more fascinated by what others will say next than you are by how you might respond.
  - Cultivate physical places that say "you matter" to others.
  - Cherish ease, feelings, appreciation, and the binary star of difference and equality.
  - Master the art of posing generative questions.

6

SPECIAL FORCES
AT NATO

*Nurture your mind with great thoughts, for you will never go
any higher than you think.*

—BENJAMIN DISRAELI

Independent thinking eludes many organizations. In its place, superficial, submissive, or domineering thinking abounds. Other organizations grasp that a culture of independent thinking unlocks each individual's unique potential at work. What will it look like for threshold leaders to prize such a culture as AI evolves?

During my career, I have coached three leaders of military special forces from three different countries. I have permission to tell the story that follows about one of these leaders whom I'll call Tony. As a coach, I'm used to maintaining confidentiality. In Tony's case, we frequently took confidentiality a step further: There are many questions I wouldn't ask, such as, "Where exactly did that happen?" Many of our fruitful coaching interactions involved minutes of silence. I include Tony's story because it shows someone who's highly successful in his field using independent thinking to integrate his whole person in the context of an organizational change program. Tony's example, as well as others you'll encounter in this chapter, point the way to addressing AI challenges in teams and organizations.

A NATO MISSION

Tony was the national special forces lead for a regional NATO mission supporting government institutions in maintaining a safe and secure environment. NATO missions are assignments carried out on behalf of defense chiefs from member countries, among others. Tony's mission safeguarded freedom of movement in relevant regions where disaffected groups frequently disturbed the peace. NATO needed to upgrade the speed and quality of its decisions in order to minimize these disturbances. In this assignment, Tony's objective was to articulate and win support for a more efficient and targeted approach to gathering information. Successful outcomes would include better senior decision-making and increased team effectiveness on the ground. Part of Tony's problem was that NATO's existing capability, structure, and processes dragged it down. He soon learned that these three things weren't fit-for-purpose and realized that the NATO mission required a major change program.

Tony takes up the story: "NATO has fostered close collaboration among its members in a way that few international institutions have managed. Conversely, being large, it can suffer from the same resistance to change and inefficiencies as other large organizations. Any given plan can stall due to political differences, mutual suspicion of others' national interests, or arguments over monetary contributions."

Eventually, Tony and his team achieved their objective of winning support for a better approach. The commander and his staff embraced the new capability. Though initially hesitant, late adopters came to see the new information-gathering approach as critical to success.

How did Tony and his team accomplish this? They focused on enhancing how others think. One way they did this was to create space for themselves and the mission to consider what matters most, which relates to the first pathway of cultivating stillness. As Tony explained about his own journey: "As a leader, this experience taught me a lot about the need for wisdom. In more complex situations, you have to look deeper [than existing knowledge and evidence]. When the fog comes in, we look to our fundamental values for guidance." By looking to his values in this new context, Tony uncovered something more of his true self. Organizationally, Tony and his team applied a resource similar to Rooted Values, encouraging purposeful reflection on what mattered most to the mission. As we will see throughout this book, different threshold leadership pathways interweave like a colorful tapestry.

Crucially, Tony introduced two elements of culture to nurture independent thinking in the NATO mission: an attentive culture and a generative culture. Before exploring each of these cultural elements in more detail, we turn first to two things that can get in the way of establishing such cultures: our response to ratchet targets and oversimplification.

THE CHALLENGE OF RESPONDING TO RATCHET TARGETS

If a culture of independent thinking matters, why don't more leaders nurture this in their organizations?

Well, being a threshold leader isn't easy. Most large or fast-growing organizations are driven by investor milestones, public targets, and/or shareholder value maximization. Like a ratchet, such targets tend to "lock in" once achieved. Typically, for the next time period, the adjusted target "prices in" what you previously achieved, requiring something more next time around. This can be a good thing when the target is beneficial. However, in an AI context, many organizations not only do not reward threshold leadership, but often actively discourage it.

Take this practical example: In November 2021, Parag Agrawal replaced Jack Dorsey as Twitter CEO. Twitter's board then set usage targets for Agrawal. How much room did he really have to be the threshold leader we needed him to be, even before Elon Musk offered to buy Twitter, and a dispute ensued? Like other tech CEOs, if Agrawal didn't meet quarterly targets for several quarters in a row, or if the stock price were to fall below a certain level, he may well have lost his job. That's the reality many executives deal with. A good way for Agrawal to meet his particular usage targets could have been to stay far from the threshold, to make us all use Twitter way more than is useful for us and to get the numbers up. An alternative (threshold) starting point would be: What level of social media or Twitter use is actually useful for us as a society? What are the checks and balances that need to be in place to fact-check and stop misinformation, some of which will inevitably affect the revenue targets? Perhaps Agrawal was near the threshold at the start of his tenure, but it is difficult to be there and stay there.[1]

This is not a personal or a company-specific observation—just the reality of the system in which leading executives operate. Therefore, a key issue we need to address in the context of technology companies is how to create beneficial AI and how to explore the boundary with AI, without encouraging or allowing executives to take every single opportunity to boost their bottom line. At the root of this issue are deeply human, empathetic factors that often need to be highlighted.

THE SCOURGE OF OVERSIMPLIFICATION

There's another reason some leaders fail to nurture independent thinking cultures in their organizations: an addiction to oversimplified thinking. Oversimplified thinking occurs when we latch onto others' first words and label them uncritically with "package deal" terms such as liberal, conservative, immigrant, tree-hugger, socialist, technocrat, or billionaire before really listening to what they meant.[2] Oversimplification is a form of gaslighting, a scourge as it cuts off other people's thinking before it has a chance to bloom.

Oversimplifying reminds me of the old man by a lake in the film *The Dam Busters*. This film was released in 1954, two years before computer scientist and cognitive scientist John McCarthy first coined the term *artificial intelligence*. *The Dam Busters* tells the story of a World War II bombing raid on German dams using the "bouncing bomb" invented by Barnes Wallis and deployed by a squadron led by Wing Commander Guy Penrose Gibson.

Amid much skepticism and opposition, Wallis persuaded the British military to use the bouncing bomb to take out large dams in the industrial region of Ruhr Valley, an area vital to Germany's war effort. Gibson assembled a new squadron with secret orders. During training, Wallis stipulated that the bombers had to fly precisely 150 feet above ground level, at just the speed he specified, and drop the bombs exactly 600 yards from the target . . . at night!

The first few times that Wallis tested life-sized bombs, they shattered on impact and the tests failed. He asked Gibson whether his squadron could fly not at 150 feet above ground level (which was already dangerously low), but at only 60 feet above ground level. Gibson agreed.

In the film, you see footage of the bombers flying 60 feet above a lake.

Then the scene cuts to an old man and his wife sitting in their cottage. You hear a succession of planes flying overhead, making such a loud noise that everything shakes in the cottage, including the half pint of beer on the man's writing desk. The man is reading out a letter he's just finished writing.

"Sir," he reads, "as a poultry farmer doing his best in the food crisis, I wish to protest against the stupid young men who indulge in idiotic joyriding at all hours. It may be good fun for them [another low-flying plane roars over the house and he looks up angrily at the ceiling] but I would point out that every time they come over my poultry houses, my hens lay premature eggs that drop off the perches and mess up the floor. This means a serious loss to both me and the country." He signs the letter, shakes with anger, and you can almost see his blood pressure rising.

Of course, the poultry farmer wasn't to know that the noise and annoying vibrations were crucial to a war effort that could help save his country. He oversimplified by making the untrue assumption that the pilots were joyriding. He latched onto a package deal phrase. He knew there was a war on and was triggered into anger, understandable given the stress of the situation.

This story shows how wrong we can be if we oversimplify and make poor assumptions. It is sad that this is exactly what often happens when it comes to AI. Ideas about AI futures are frequently sold as package deals, such as "singularity," "technocalypse," or "Superintelligence." Sometimes it feels like a war rages over how we should develop technology. Sometimes, it feels like a war rages over our concept of being human. Those who take the time to be independently thoughtful are less likely to fall into oversimplified or package deal traps. Organizations that resist oversimplification remain true to themselves in an evolving world that is constantly trying to make them something else. The way to overcome these habits is to create generative, attentive, safe settings for your colleagues to think *for* themselves, *as* themselves. This matters because, as Professor Fei-Fei Li, codirector of the Stanford Institute for Human-Centered Artificial Intelligence, pointed out, advanced AI requires refined human thinking.[3] Part of this refinement is coherence and nuance, compared to oversimplification. At the threshold, leaders will refine culture around them in attentive, generative ways.

ATTENTIVE CULTURE

In an attentive culture, team members cultivate fascination with the direction others will go next in their thinking. Tony astonished skeptics and early adopters alike by inviting them to meetings in which he was genuinely interested in what they would say next. His NATO team then followed suit. Together, they managed to avoid the adrenaline of interruption and resist what Jason Fried calls "Refutation Mode," in which there is no listening. As Nancy Kline put it, "attention generates thinking."[4] In this way, Tony and his team generated higher quality thinking about their objective.

"Giving attention to early adopters increased their ownership of the idea so they would start to champion the change," Tony said. "Listening well to the majority stopped them becoming laggards and eased in new processes. Personally, I hold to the idea of trying to 'love your neighbor as yourself.' This was not always easy, though, especially when dealing with people who had a Kafka-esque approach to process. Fighting through tortuous bureaucracy, with people creating barriers but dodging responsibility, would certainly cloud any feeling of love I had for an individual!"

Tony's words about ownership matter greatly. When people reach their own insights and conclusions, solve their own problems, or come up with their own ideas, their brain gets a hit of the pleasure chemical dopamine, which helps facilitate action.[5]

It must have been tempting to hurry up the laggards. Instead, Tony shunned silent or vocal put-downs, in the spirit of the following quotation:

> A new idea is delicate. It can be killed by a sneer or a yawn; it can be stabbed to death by a quip and worried to death by a frown on the right man's brow.[6]

Centered, Tony demonstrated threshold leadership. He encouraged NATO to pay magnificent attention to what it was doing. As a result, early adopters stepped forward to shape the change program, buy into it, and ultimately champion it. This program affected systems, processes, training, and other key organizational ingredients. In this way, Tony's approach touched organizational culture, not just individual leadership mindsets. I later coached Tony using the Thinking Pairs resource. Tony remarked that attention-rich approaches like this were key to the success of this NATO mission.

THE SYSTEMIC IMPORTANCE OF ATTENTIVE CULTURES

Nurturing an attentive culture matters for systemic reasons, not just for individual organizational initiatives or teams. One of the strengths of Tony's approach was the thoughtful way in which he created the conditions for senior leaders to challenge more complex, system-wide assumptions, and to embrace transition. Such thoughtfulness and challenge matters, of course, in relation to NATO's intergovernmental military priorities. But it also matters in relation to other international priorities such as reducing food waste or appropriately responding to the rising average temperature of our Earth's climate system, areas in which AI is taking an increasingly prominent role.[7]

During a discussion about governments and AI, Kay Firth-Butterfield, head of AI and machine learning at the World Economic Forum's Center for the Fourth Industrial Revolution, said, "We're creating iterative, agile governance around a technology that is in itself changing almost as frequently as we think about it."[8] In a world where AI and its context are changing fast, threshold leaders will protect time to think creatively about vision, context, and assumptions in non-interruptive, generative ways.

At the threshold, leaders heed the call to spend more time in transition. This can be hard, when increasing specialization leads to institutionalization and bureaucracy, with the diminution in transitory spaces that accompanies this journey. I have seen formerly nimble start-ups get bogged down in stifling process, when leaders refuse the call to attend to what is really going on around them, as they did in the early days. What if transition became your permanent condition?

Leaders who nurture cultures of independent thinking also challenge the bystander effect, which first came to prominence after the murder of Catherine "Kitty" Genovese. On the morning of March 13, 1964, Genovese, a twenty-eight-year-old bartender, was stabbed fourteen times outside her Queens apartment in New York City. Two weeks later, the *New York Times* reported that thirty-eight bystanders saw or heard the attack, "turn[ing] their back on Genovese's early morning cries for help, shutting their doors to silence her screams."[9] The bystander effect holds that individuals are less likely to intervene in an emergency, the more witnesses there are. As AI witnesses more of our lives, what will you assume about your role as a leader? Will you become more of a bystander? Threshold leaders work to minimize this systemically important effect. The more you think for yourself, the more you will access what really matters to you, even if what matters changes, and the less you will become a bystander.

ATTENTION AND PSYCHOLOGICAL SAFETY

These observations about systems and bystanders remind me of my journey at Coachify.[10] In my final months as CEO, I had become less inspired by what I was doing. I had slipped into mindless obedience to a distant "ideal" of raising capital and driving financial value. I was no longer doing my finest thinking and this translated into lower impact, as customer feedback and financial results testified. I banked out of this tailspin by having Thinking Pairs with others within and beyond my immediate team. The attention and ambition that my thinking partners showed me caused my joy to return and our culture to become more attentive. I felt alive and effective, and I generated my finest thinking in years.

One of the most important achievements of a thinking partnership—at NATO, at Coachify, and elsewhere—is establishing psychological safety, which refers among other things to team members not being afraid to express themselves as they are and as they want. In its quest to build the highest performing teams, Google found that psychological safety matters most.[11] As AI dominates more work processes, employees may fear for their security and even their value in society. In such moments, psychological safety will matter even more.

Directive thinking destroys psychological safety. I was once in a meeting where someone cried and the meeting convenor seemed uncomfortable as a result. The convenor immediately called a halt to the meeting, suggesting that we continue later. However, at no point had the person who cried indicated a desire for a break. "Being sensitive to others' feelings" can easily slip into "deciding for others how they should continue." The bond we were building as a group dissipated, and it took days for our shared thinking to approach its previous level.

By contrast, a culture of appropriate emotional release restores thinking to groups and organizations. As Nancy Kline put it, "Listening through anger makes way for thorough thinking, crying can make you smarter and after laughter thinking improves."[12] Once you see the value of paying magnificent attention, you can't unsee it. This threshold way of leading taps into uniquely human qualities of safety that will make our organizations effective, responsive, and vital as AI accelerates.

One organization that can't unsee this is Janssen (part of Johnson & Johnson). Over the last few years, I have been part of a team helping Janssen embrace Thinking Environment* principles. Pre-COVID, if you walked into any Janssen meeting room in the United Kingdom, you would likely see physical reminders of the importance of paying attention and nurturing one another's finest thinking. Janssen relies on a bedrock of mature leadership principles, and these principles helped it build ownership and psychological safety, and respond well to one of humanity's greatest health crises.

When I spoke about this with Simon White, head of learning and development at Janssen UK, he pointed to the compassionate leadership shown throughout by Johnson & Johnson during the pandemic. Said White, "Those who create thinking environments benefit from increased resilience and innovation, such as adopting new practices that support product breakthroughs." At Janssen from March 2020, teams came together every few days for half an hour, using Thinking Environment* principles. For example, they took time to hear from one another about their nonwork lives by using rounds.[13] Strategy design meetings now regularly incorporate intentional space for others to think. White noted that these features have been "hugely beneficial." This kind of culture created the conditions for inclusive, incisive thinking throughout Janssen and Johnson & Johnson.

Attention is a universal, strong, and glorious force. As you encourage a culture of attention and generative questions, you can expand the boundaries of what your teams and organization can achieve in the Age of AI. What will it take for you to be this kind of leader?

GENERATIVE CULTURE

During the NATO mission, Tony and his team also created a generative culture, which is a culture that encourages everyone to generate their finest thinking. Tony contributed to this by focusing on three areas: generative questions, generative talent, and generative processes.

First, Tony role-modeled, repeatedly asking generative questions during important meetings. He explained what generative questions are and inspired others to use them. He knew that asking courageous, generative questions was an intelligent thing to do.

Second, Tony set up his team for success by upskilling them to get under the skin of assumptions, for example, by asking, "What are you assuming that is stopping you from achieving your part of the project goal?" The payoff for leaders in the Age of AI is not just individual and organizational, it is social. Philosophers and business consultants have found that leaders who operate in healthy, assumption-forming environments formulate their thinking with greater interpersonal intelligence.[14] In turn, this yields more energy for action and better outcomes.[15] To enjoy these benefits, use the resource Digital Thinking Transformation at the end of this chapter.

Crucially, Tony's colleagues explored assumptions about difference. NATO is functionally, professionally, and ethnically diverse, to name just a few dimensions of complexity. Wide-ranging and inconsistently held assumptions abound at NATO as elsewhere. In Tony's case, a key success factor was establishing a shared understanding of the value of different perspectives and backgrounds to the mission. Difference is a key part of a generative culture.

Third, Tony adopted impact-oriented processes that supported a generative culture. For example, he applied a resource similar to the Pre-mortem Bias Reducer, so that his team could generate even more effective thinking. Tony also regularly gave positive feedback to team members who helped others remove assumptions that blocked progress.

Generative cultures can raise profit and productivity. Emily Havers and Beverly Whitehead conducted research into the impact of an organization nurturing a Thinking Environment*. One company managing director noted that "these meetings have produced the best results of any meetings

I've been in." A regional director in financial services reported that business "has improved by at least 20 percent, and that's measurable in financial terms." Other benefits noted in this research were more productive working relationships and faster resolutions.[16]

For a generative culture truly to take root, impact must also be viewed through a long lens. Tony gave others the heart to go to the unexplored edge of their ideas, knowing that it is at these edges that valuable progress is often made.

Consider these words by author and executive coach Trevor Waldock:

> The most unselfish thing you can do with your life is to plant a walnut tree. If you plant a walnut tree, you won't see its fruit for many years—you're investing for your children. You are planting something for generations to come.[17]

At the end of your life, will you be able to say, "I planted a walnut tree"? Will you be able to say that you inspired your teams and organization to think well about the challenges and opportunities technology provides?

As technology advances, let's invite deeply human, empathetic factors. Leaders at the threshold encourage generative, attentive cultures that contain these factors. Such cultures foster a connection of thinking and being in organizations, which will result in wiser planning, more inspiring strategies, more energized people, and more effective execution.

THRESHOLD RESOURCES

RESOURCES FOR LEADERS IN LARGE ORGANIZATIONS

Resource 9. Nudges Toward an Attentive, Generative Culture
PURPOSE
Encourage your organization to move in an attentive, generative direction.

PROCESS
Use these lists of inputs and questions as sources of inspiration when you are planning meetings, reviews, or projects.

I.   NUDGES TOWARD AN ATTENTIVE CULTURE

Here are four ways to encourage your team to bring more of themselves to their thinking:

- Trust your team's intelligence even as AI provides more solutions. For example, even if you are more senior than a colleague, your colleague's experience may be more relevant to a particular machine-learning challenge you are facing. Be attentive to colleagues' thinking, feelings, and intuitions.
- Tell your team that expressing feelings is part of good thinking. For example, encourage your team to be present to one another's feelings about AI and other topics, rather than seek to escape their feelings.
- Tell your team that diversity adds quality to thinking. Encourage each person to draw on his or her own experiences, background, perspectives, and other aspects of who they are, as far as they think relevant. Avoid the false assumption that your team is homogenous, as this assumption introduces unreality to discussions, and this unreality lowers the quality of those discussions.
- Don't assume that you can do others' thinking for them better than they can, just because they are crying.

II.   NUDGES TOWARD A CULTURE OF GENERATIVE QUESTIONS

Try socializing some of these higher-level questions throughout your organization:

- What will we learn about AI in a year that we already know?
- What might we be assuming here that is limiting our thinking on this issue?
- What would need to be true for our company to become the market leader in AI in our industry? (Or if we are already the market leader: What would need to be true to double our impact?)
- What if we encourage our closest competitor? What could this look like practically?
- If you were the CEO, what problem would you solve first and how would you do it?

- What are we assuming that is stopping us from evolving our corporate governance for an AI age?[18]
- What are we assuming about ethics that is stopping us from figuring out our values?[19]

I also encourage you to develop your own fit-for-purpose generative questions.

III. NUDGES TOWARD A GENERATIVE CULTURE OF LEGACY
Inspire others to articulate their situation and vision in relation to legacy, using four groups of questions. Each group corresponds to one of Trevor Waldock's leadership stages:[20]

1. Experimentation: What can you uniquely do as an AI leader? What can you learn from what you can do?
2. Experience accumulation: What do you need to do to move from unconscious incompetence about AI toward conscious competence?
3. Effective leadership: In this stage, you stop asking tactical questions like "What do I need to do to fit in around here?" "How do I get approval for my AI ideas?" and "How do I get promoted?" Rather, you begin to ask, "What is the difference that I want to make in the world?" and "What is the legacy I want to leave to those who come after me?"
4. Eldership: A more communal approach where you focus more on what is necessary to help others flourish and achieve lasting impact. Through such an approach, you may create a life-giving future with better outcomes, rather than just react to what's going on. Through eldership, you help others connect their thinking and being and thereby express more of what makes them magnificently human. Ask yourself: If you knew that you are planning walnut trees, what would change?

IV. NUDGES TOWARD A GENERATIVE CULTURE OF TALENT
Use the following questions to increase the chances of success when you set up a talent pool composed of humans and machines:[21]

- AI assistants: What are you assuming about assistants? What are you assuming that stops you from training the assistants further?
- AI monitors and monitors: What are you assuming about real-time feedback? What are you assuming that stops you from identifying discrepancies? What are you assuming that makes you limit the scope of your AI rollout?
- AI teammates: What are you assuming about how humans and machines relate?

In an AI context, "team" or "talent" can refer to humans, AI, and a hybrid of both.

PAYOFF

Consistently invite the finest human qualities into the heart of your organization.

Resource 10. Digital Thinking Transformation

PURPOSE

Exploit knowledge as you pursue digital- and/or AI-related transformation.

PROCESS

A three-step thought starter for threshold leaders to raise the independence and therefore quality of a team's thinking about a digital- or AI-related change. The change could be toward shifting your digital business model, prototyping an idea, or designing another digital or AI-related initiative.

This resource draws on Roger Martin's Knowledge Funnel; Rita McGrath's Discovery-Driven Planning, helped forward by Ryan McManus; and Scott Keller and Bill Schaninger's approach to innovation and learning.[22]

AI affects organizations unequally, so shifts that different leaders target will naturally differ by organization size, stage, sector, and geography. Digital- and/or AI-related transformations are therefore complex topics. This resource is also, emphatically, not a full change process. Consider it more of a spark that might help ignite a change process, an abbreviated thought starter rather than a comprehensive guide.

PART A. GET STARTED WITH OBJECTIVES AND OUTCOMES

Define what transformation objectives you want to achieve by identifying what isn't fully working in your digital- or AI-related operation. Some companies take an analytical approach to this step. Others take a Design Thinking approach, which is a way of reasoning especially suited to engaging with design problems. Which of these you choose depends in part on your existing culture of strategy development. Encourage all involved to be fascinated by what others think about objectives.

Assess where you are compared to your objectives, and then identify and prioritize outcomes and metrics as follows:

- Ask yourselves: What do we value most: better, faster, cheaper, or more convenient? Where do we think technology could add the most value? If we knew we could serve humanity through our work, what would change?
- Define clear targets that will guide you as you learn. For each objective identified above, what would success look like, and by when? Some targets may be project-specific, such as reducing the time it takes sales employees to reach 90 percent of their target from ninety days to thirty-five days within a year. Other targets may be organization-wide, such as return on time invested.[23] Scrutinize these targets against what matters most (to you personally, to your organization, to society) in the long term.

PART B. ITERATIVELY EXPLOIT KNOWLEDGE VIA A LEARNING ENGINE

Iteratively exploit knowledge that you have and can gain in relation to your emerging outcomes. Start this process as early as possible, and ideally in parallel with some of Part A above. You might proceed as follows:

Empathize: Learn more about what matters to your customers, digitally and in relation to AI. Take as much care as you can to avoid bias in your selection of and interactions with customers. Synthesize this knowledge. Remain humble and curious as you take others' input.

Define: Define customer needs that will be solved by your AI and digital transformation.

Learn: Learn from and encourage your "competition."

- Identifying your competition is not a simple step, as traditional industry boundaries have blurred in recent decades. McGrath and McManus propose a strategic arena approach to this step: Think about what your customers want to get done, then consider who else may be providing that. For example, "Netflix has been very clear that it doesn't intend to compete just against television or the movies for viewers' time. It intends to compete against every possible leisure activity that a person might do instead of watching streaming content [including] . . . magazines, books, podcasts, and sporting events."[24]
- Reflect on what you can learn from your competitors. What are their digital strengths and weaknesses? What are they not doing that you could do? What if you combined the AI strengths of one with the AI strengths of another?
- Consider how your sector, your customers, and your other stakeholders may benefit if you collaborated more deeply with competitors than is traditionally understood.
- You may need to return to and refine the "getting started" stage after you have identified competitors and collaborators.

Ideate: Come up with as many creative solutions as possible, drawing on your full breadth and depth of talent pools. As part of this step, encourage your team to consider and explore the boundary between humanity and machines.

Prototype Your Idea: Remember that this is just a rough draft and that this approach to shifting your digital business model is incremental. You are still at the exploration stage.

Test: Share your prototyped idea with customers for feedback. What worked? What didn't? Scrutinize the economic impact of the prototype, if scaled under various growth scenarios.

PART C. ENSURE SUCCESS VIA REFLECTION

- Draw together learnings from your pilots and experiments.
- Distill the wider learning implications for the rest of your organization.

- Ask, "What are we missing?" Challenge your team to come up with further generative questions, as generative questions prompt the highest quality thinking.
- Throughout the digital transformation process, encourage senior leaders and other coworkers from throughout the organization to be involved, to enable broad engagement. Try to avoid over-directing the process. You will assemble a more engaged coalition the more they sense that you created an environment for them to think well about the initiative.

PAYOFF

Hurdle common barriers to digital- or AI-related transformation.

CHAPTER TAKEAWAYS

At the threshold, leaders build attentive and generative cultures, which are characterized by psychological safety, nuance, engaged talent, and intentional legacy.

- Such cultures matter systemically in the Age of AI.
- Our natural response to ratchet-like targets makes achieving such cultures hard.
- Transcend this difficulty by taking a long-term view, helping others travel to uncharted parts of what is possible.

EMBODYING INTELLIGENCE

7

DINNER WITH FRANCIS CRICK . . . OR JAMES WATSON?

The fact that machines are not made of flesh makes more of a difference than we realize.

—INSPIRED BY LARISSA MACFARQUHAR[1]

In February 1997, when I was a third-year undergraduate, I hosted a dinner that I now wince to recall. I was serving as president of the Cambridge Union Society, the oldest continuously running debating society in the world. Especially for a foodie and (usually) conversationalist like me, one of the delightful aspects of being president was getting to dine with famous guests before debates or lectures. One evening at a local restaurant, I hosted Francis Crick, the British molecular biologist who, with American molecular biologist James Watson, coauthored an academic paper proposing the double-helix structure of the DNA molecule. Or rather—and this is the embarrassing part—I dined with *either* Francis Crick *or* James Watson, and I couldn't remember whom.

Here was the British (or maybe he was the American) molecular biologist who had coauthored one of the most significant academic papers of the century. I was in the presence of a Nobel Prize winner! There was so much to learn from such a man, so much to ask. And yet my curiosity was lacking. I simply hadn't prepared for the dinner by finding out more about him. So, the conversation at my end of the table was somewhat stilted. Eventually another guest (who was sitting two places down and who was interested in the natural sciences) kicked me under the table and suggested we swap places. I fixedly stayed in my seat for a while before eventually moving. (I'm still wincing as I write this.)

I think I know why the identity of my dinner guest hadn't stuck in my mind. Usually, I would be riveted and engaged on such an occasion. I wasn't thinking well, because I was physically, mentally, and emotionally shattered. Physically, I was in poor shape. Youth can mask a lot. But late nights and regular large, boozy dinners with scant recovery meant that I'd fall into bed exhausted and had little energy in my discretionary time to invest in other, more creative tasks, such as finding out about well-known dinner guests. On top of this, I was studying hard, fulfilling my Union duties, and more. I've always been a highly focused person, sometimes too much. I had pushed myself so hard for two years that I was mentally glazed, to the extent that I didn't even retain information about a famous person who had been seated right next to me![2] My exhaustion also fueled harmful emotional reactivity. That night in Cambridge, I didn't care enough about how welcome my guest might feel to get interested in him. I embodied lack of interest.

By contrast, threshold leaders embody intelligence. Path III explores what embodied intelligence is, why it matters for individual leaders in the Age of AI, and how you can become more intelligent by working with your body. The core idea is: The fact that machines are not made of flesh makes more of a difference than we realize.

EMBODIED INTELLIGENCE

In individuals, embodied intelligence refers to the brain adjusting to feedback from its body and other parts of its surroundings.[3] Embodied intelligence includes knowing how to recover energy, as well as how to use physical movement in inspirational, effective ways. As part of this, an embodied leader prizes the simple experience of *being a body*, in social relationships with others. This simple-sounding idea turns out to be revolutionary. In the words of author Hillary McBride, embodiment "is to be present to yourself and your experience from the inside out."[4]

Here, we move away from a purely cerebral form of leadership, as if logic or cognition could solve everything.[5] Embodied intelligence takes you toward an integrated view of humanity, rather than a dualistic one that views the body as separate from the brain or mind. This integration matters. AI won't match human embodied intelligence because of the nature of our felt bodily experience.[6]

To those who have suffered physical abuse, trauma, or other pain, embodiment may feel unsafe. For example, some leaders do not feel happy with how their bodies look or move. If you recognize any of this as part of your story—as victim, perpetrator, or both—please be assured that I am not trying to offer simple answers, as if such answers exist.

If you are recovering from trauma and want to find a way to engage with embodiment, there is no substitute for seeking out a trusted trauma therapist. For example, you may wish to explore body-based therapies such as sensorimotor psychotherapy or somatic experiencing.[7] Such therapies can help you learn from past experiences and explore what moving forward means, given that these experiences are part of your past and present. You might also consider treating my four pathways as gentle invitations, seeds from which life may grow.

TRENDS

Three current trends show the need for embodied intelligence: a rising tendency to prize overly rational thinking, increasing pressure to mimic machines, and encroaching techno-humanism.

First, many leaders in the West unhelpfully identify themselves exclusively with their rational thoughts, an identification that goes as far as back as classical Greek philosophy. Around 375 BCE, the Greek philosopher Plato presented his Allegory of the Cave in his Socratic dialogue, *Republic*. In this allegory, prisoners see shadows (who denote what is real), but not the thing that is making the shadow (which denotes what is ideal). From this, Plato's idea arose of the body as the tomb of the soul (or mind), as well as its tool.[8] This is a dualistic view in which your body isn't so much part of you as merely an instrument you use to walk around in.

Fast-forward to the Age of Enlightenment, a European intellectual movement of the seventeenth and eighteenth centuries. In 1637, the French philosopher René Descartes coined the phrase *cogito ergo sum* in his *Discourse on Method*. This phrase translates as "I think, therefore I am" and conveys the ideas that our real self is our ability to reason our way into truth and that our bodies are somehow separate. A related idea is that our bodies get in the way of what really matters: thinking. These notions pervade many fields of AI development.

Today, classical and Enlightenment philosophies influence leaders in many positive ways, such as via the Socratic method of dialogue. However, when it comes to embodiment, many leaders embrace a dualism that unhelpfully separates body from mind. Author and mystic Cynthia Bourgeault noted that disembodied thinking creates "a superficially tidy universe [where] everything is in a box. It is clean. It is just not real."[9] In other words, a dualistic separation of mind and body tends toward unreality. Against this, threshold leaders shift from "I think, therefore I am" to "I think with all I am." They jettison overly narrow approaches and adopt holistic ones in which they are fully connected to their bodies and therefore fully alive.

Second, leaders face increasing pressure toward disembodiment, as they face rising pressure to mimic machines. This pressure has physical, mental, and emotional effects.

I regularly encounter leaders whose physical energy is chronically low. This has the effect of putting dampers on their thinking, as they are too overwhelmed or disconnected from their bodies to bring their whole self to bear. We didn't need decades of research to tell us that sleep matters. Shakespeare summarized it beautifully in 1611 in *Macbeth*, act two, scene two, stating that sleep is "the chief nourisher in life's feast." In contrast, sleeping an hour or two per night under a desk remains a cherished rite of passage for many entrepreneurs.

In terms of mental pressure, we know that cognitive multitasking doesn't work (for humans), but leaders who direct billions of dollars and influence billions of lives still text during board meetings. Some executives expect work colleagues to respond to a message within an hour during the weekend as if it's possible to be always on like computers and/or to manage by belittling. Others look on and copy.

Tony Schwartz, author of *The Way We're Working Isn't Working*, wrote in *Forbes* magazine in 2020, "How you feel so profoundly influences those you lead." Yet many leaders lead robotically by failing to nurture positive emotions and by denying their anxieties. Where this failure occurs, organizations are endangered, exposed, and ineffective.

Third, techno-humanism is encroaching on us and threatens our access to embodied wisdom. According to historian Yuval Noah Harari, techno-humanism is a religion that seeks to use technology to improve humanity to godlike levels. I don't view techno-humanism as necessarily religious, but with sophisticated implanted technology it is not hard to picture a future techno-human race where artificial and human "software" and "hardware" merge. For example, imagine a world of cloud-enabled synthetic neocortices fused onto or inside our skulls, allowing intelligent uploads and downloads. Today, of course, a chasm separates theory from practice in this area. But if we ratchet down this thought experiment a couple of notches, we can see that a more limited form of techno-humanism already dwells with us.

Just after Dana Hanna took a solemn vow to love and cherish his new wife for the rest of his life, he took out his phone, updated his relationship status on Facebook, and tweeted a wedding announcement. These are unprecedented times. As a society, we are hurtling headlong toward ever greater AI-fueled digital dependence, and there's no road map. What's the difference between an AI-enabled brain implant and always having your phone in your pocket? The embedding of technology in our bodies is not far off and has already started in some quarters, but for the time being, we keep our devices constantly at hand.

The advantages to this are clear and multiple—we're connected, informed, challenged, and entertained like never before. Our voices can have a reach unimaginable to our forebears. Accessing the thoughts of the world's most original thinkers, starting innovative new businesses, and finding the answer to almost any question in seconds is now commonplace. Some might argue that the onset of techno-humanism offers signs of hope for leaders using disembodied thinking.

But poorly led, techno-humanism comes at a cost. The dopamine-fueled draw of our devices is already significant, even addictive, and it pulls us away from real, embodied, human interaction. It is sad that AI is already with us in ways that threaten to downplay our access to the wisdom our bodies provide. When we become dependent on a regular digital hit, we go shallow, across a scattering of computer-generated quick-reads and sound bites. Our performance dips, our focus fragments, our creativity fades, our real-life relationships suffer, and our mental health plummets.

DEVELOPING EMBODIED INTELLIGENCE

Threshold leaders embody intelligence at two levels: energetically and via multiple intelligences.

Findings about leadership and energy are not new. More than two decades ago, the *Harvard Business Review* published an article by Dr. Jim Loehr and Tony Schwartz in which the authors argued that recovering emotional, mental, physical, and spiritual energy is crucial for performance.[10] Nor was this article the first published work on the topic of energy and leadership.[11] In the last few years, managing energy has become not just professionally mainstream but also clinically advisable. Thousands of peer-reviewed studies have linked poor sleep to pretty much every health problem you can imagine, and the COVID-19 pandemic has thrown the importance of home and workplace well-being into sharp relief. Many smart leaders remain seduced by machinelike work strategies. Managing human energy will be critical in the Age of AI, and most leaders don't do it.

The best codification of the four dimensions of energy that I have encountered is in the book *The Way We're Working Isn't Working: The Four Forgotten Needs That Energize Great Performance* by Tony Schwartz, Jean Gomes, and Dr. Catherine McCarthy.[12] This book brims with insight and useful application. Some years after I first read this book, the opportunity arose to meet and work with Jean Gomes and some of his team in Europe. I have also found the work of Mark Oakley and Matthew Walker extremely insightful on the topics of human rhythms and sleep, respectively.[13]

Schwartz, Gomes, and McCarthy argue that leaders excel the more they pulse between performance and recovery in the following four dimensions of energy:

- Spiritual: pulse between nurturing others and nurturing self.
- Emotional: pulse between high and low energy, remaining positive in both states.
- Mental: pulse between big picture and narrow focus.
- Physical: pulse between recovery and expending energy.[14]

I view these four dimensions as the foundational level of embodied intelligence. Working well with these four dimensions matters in the Age of AI, because unless leaders act, AI could further harm our mental well-being and presence.

Beyond this, threshold leaders access another level of embodied intelligence. They capitalize on poetic and kinesthetic intelligences, two of psychologist Howard Gardner's multiple intelligences, which we encountered in Path I. Poetic intelligence involves using breathing and the physicality of emotions to foster poetic leadership. This approach draws on the poetic aspect of Gardner's linguistic intelligence, defined as "finding the right words to express what you mean." Kinesthetic intelligence may be defined as coordinating your mind with your body and involves using knowledge that resides in your body.[15]

We may therefore summarize the two levels of embodied intelligence as follows:

1. Foundational level: Managing your energy—drawing on well-known findings about leadership performance
   - Proper selfishness
   - Emotional performance
   - Mental focus
   - Physical freshness
2. Advanced level: Putting it all together—drawing on aspects of multiple intelligence
   - Poetic intelligence
   - Kinesthetic intelligence

The rest of this chapter and the next explores these levels predominantly in an individual leader context. Chapter 9, the final chapter in Path III, explores embodied intelligence in a team and organizational context.

PROPER SELFISHNESS

A friend told me the following story about Marines helping out after an earthquake in South America in the 1980s. After weeks of arduous physical relief work, one squad returned to their ship moored off the coast. Exhausted yet desperate to continue helping the local population, the squad leader said to his commanding officer, "I need to get back out there, *now!*"

The officer calmly surveyed his colleague and responded: "No, you need to take rest. You are putting your own men in danger. Show some proper selfishness."

I don't know whether the above story constitutes the first use of the term *proper selfishness*. This term took hold in the military, and Charles Handy wrote about it in his book *The Hungry Spirit*.[16] In one sense, we are all familiar with the idea. *Put on your own oxygen mask before helping others.* If you haven't recovered enough to take care of yourself a little, you're no good to anyone else.

I regularly notice leaders failing to nurture themselves because of an untrue assumption about what counts as selfishness. Schwartz and Gomes frame it beautifully: "Selfishness is about putting yourself first at the expense of others. Taking care of yourself is a critical ingredient in being able to take care of others."[17] But some leaders try to serve their team, company, and family while failing to take care of themselves. Taking enough care of yourself is proper selfishness.

At its heart, proper selfishness is not about merely taking physical care of yourself. It's about nurturing your values and purposeful foundation as fuel to nurture others. Proper selfishness encodes two ideas: First, take care of yourself because, second, only then may you take care of others. As part of this, the more strongly you ground yourself in beneficial purpose, the better you will be able to serve others and the more sustainable will be your success.

One of the most gripping business books I have ever read is *The Smartest Guys in the Room*, which charts Enron's journey from its origins to a financial peak reaching a valuation of $70 billion in the year 2000 to collapse in a cloud of scandal.[18] A key reason Enron's executives fell short was that they relentlessly avoided reflecting on beneficial meaning and purpose. Novelist Richard Powers noted that meaning is found in balance, as is the case with light and darkness, death and life, and contraction and growth.[19] Selfish growth, consumption, and self-obliterating service are poor goals for humans. The contraction and growth throughout our bodies stands as a metaphor for the pulsing that we do at our best.

The alternative to taking care of yourself and others is harming self and harming others.

We may link this observation with the Stroop task, which is one of the best-known psychological experiments, named for the psychologist John Ridley Stroop.[20] In this task, subjects become much slower at naming a color when there is a mismatch between the ink color and the meaning of the word (such as when the word *green* is printed in red ink). What is going on here is that when a word is dissociated from its semantic meaning, subjects in the experiment slow down. We can use the Stroop results to help us think about core identity and meaning. By analogy, if we dissociate a person from his or her core identity and meaning, we add in another layer of processing, which takes effort and slows the person down. This effort uses valuable cognitive resources to do things like suppress natural responses and mechanically think through what the right actions would be. Similarly, spiritual and mental energy are linked. Enron's leaders thought they were being smart, but they had slowed themselves down by failing to connect their thinking with some important, deeply held values. Leaders who connect their thinking with meaning and purpose will thrive and enable others to thrive.

Is threshold leadership narcissistic? What if the four threshold pathways just fool us into thinking we're evolving, and in reality they are self-serving tools of self-enhancement? This is a risk, and the responsibility to avoid it lies with each of us. To avoid the risk, I encourage you to focus less on the extrinsic benefits arising from the pathways in this book, benefits such as your leadership influence or performance, and more on your awareness of the process of crossing the threshold. Practicing this awareness is in itself a good thing and invites you toward your leadership edge.

EMOTIONAL PERFORMANCE

Successful leaders go beyond grounded purpose to cultivating positive emotions, which are fuel for high performance and satisfaction. Emotional positivity does not imply a fake kind of happiness and does not absent critique: otherwise, it may be a recipe for abuse. Schwartz, Gomes, and McCarthy describe emotional positivity as high energy or low energy and may include contentment, interest, and hope, as well as cheerfulness, euphoria, and eagerness.[21] At your best, you can have profound energetic rhythms and become an embodied, feeling thinker.

In my consulting work, I have seen fire return to the eyes of men and women who formerly felt overwhelmed, lost, or empty, thanks to learning to increase their emotional performance. Emotions are relevant to nearly everything we do. In a world where the boundary between humans and machines blurs, the risk is that we will downplay our feelings, preferring instead to rely on cloud-based intelligence instead of instincts and intuition. The risk is that we get lazy at attending to important emotional cues within and around us. In an assimilated AI future, it will be much more important to attend to emotions, as things bring our intelligence to life.

Every pathway in this book connects to emotions. Threshold leaders who are emotionally high-performing cultivate stillness (Path I), create Thinking Environments® (Path II), and explore a journey of increasing maturity (Path IV). Along with the resources in Path III, these are the finest ways I know to encourage another person into a positive emotional state. By doing this, threshold leaders connect their feelings beautifully with their thinking, they let go of exhaustion, and they perform better.

THRESHOLD RESOURCES

RESOURCES FOR ANY LEADER

Resource 11. Feel What Matters Most
PURPOSE
Connect what matters most to you with bodily emotions.

PROCESS
A five-minute reflection exercise.

Either read the steps below in advance of doing the reflection or record the steps as a voice memo. If you use a voice memo, make sure to leave pauses of approximately one minute between the steps. This will help you focus fully on the reflection, as you will avoid the need to keep pausing and unpausing your digital device during the exercise.

STEPS:

- Sit quietly and contemplate one aspect of what is most important to you. This may be your purpose, your life mission, a value that matters to you as you live your life, or something else that is meaningful for you.
- Reflect on how you feel about this, bodily. As you bring to mind this thing that matters to you, where can you sense emotions in your body? How positive or negative are the feelings? How strong are the feelings?
- Choose one of the parts of your body where you identified a feeling. Imagine sending your breath there. Inhale and exhale deeply, several times, as you imagine sending your breath to that place.
- Reflect on what feels physically different now, if anything.
- Return your attention to what matters most to you. What new perspectives or patterns do you see? What new questions arise for you?

PAYOFF

- Increase your deeply felt connection to what matters most for you.
- Recognize other ways of seeing the world.
- Open up new possibilities in your leadership.

Resource 12. Elevate Your Emotions

PURPOSE

- Tap into the reality that your team embodies emotion and that emotional knowledge resides in people's bodies.
- Increase your team's emotional literacy and positivity.

PROCESS

A three-step process to be used during a team meeting or an away day. The three steps are:

- Invite
- Notice
- Appreciate

1. INVITE

    Make the following points:
    - Try to bring your best attention and your best thinking to this exercise.
    - The more you fill your reservoir full of positive emotions, the better you will bounce back from upsets.[22]
    - This exercise provides a way of filling your emotional reservoir positively.

Invite everyone to center themselves, using a mindfulness technique. For example:

Ask each person to close their eyes and notice their thoughts, without judgment.

Or: Ask each person to take a few deep breaths with their hands on their belly or chest, and notice how it feels to breathe. (Noticing breathing is important as, after all, all living beings are breathing.)

2. NOTICE

    As appropriate, ask what people notice now. Note that our breath is critical to elevating emotions. This is especially true in our relationships. As Hillary McBride wrote, "The person you come to see as your hero or your enemy took a breath right now, just as you did."[23]

    In many languages, there are thousands of words that label emotions. For simplicity, here are twenty-six positive emotion words taken from the website www.positivepsychology.com:[24]

    Joy
    Elevation
    Enthusiasm
    Gratitude
    Altruism
    Eagerness
    Serenity
    Satisfaction
    Euphoria
    Interest

Relief

Contentment

Hope

Affection

Enjoyment

Pride

Cheerfulness

Optimism

Amusement

Positive Surprise

Happiness

Inspiration

Confidence

Love

Awe

Admiration

Share these words with the group, ideally without using a screen. One way to do this is to write up the words in advance on a flip chart.

Ask each person to do the following:

- Bring to mind three of the positive emotion words (or some other positive emotion words if they prefer).
- Consider what physical sensations typically accompany each of the three positive emotion words. For example, smiling may accompany joy, an upright posture may accompany confidence, and a churning stomach or fidgeting toes may accompany love.
- Bring to mind times in your life when you most need these positive emotions. For example, you might notice that you need altruism when a negative emotion such as annoyance blocks you from engaging lovingly. This "noticing" can help you shift perspective and open up new possibilities other than your usual, default responses.
- Consider when during the next week you can tap into these positive emotions. Be as specific as possible. For example, you may choose to tap into a particular positive emotion every time

a colleague triggers you. Or you may choose to remind yourself of a particular positive emotion at the start of each day. In any case, be intentional about when you will tap into the emotions you named.

- Each evening, write an entry in a gratitude journal. What you write could take the form "I feel [___] because . . ." (Fill in the first blank with a positive emotion for each line you write.)

3. APPRECIATE

As you close the meeting, ask each person in the team to appreciate a quality in another person. This step brings positive emotions to life and allows you to role-model elevating emotions. You might quote recent research that suggests that optimists have a better chance of living to age eighty-five or older.[25]

- Start by asking each person to write down one quality that they appreciate about each of the other team members. "What I appreciate about you is . . ." Often, you know these qualities deep down, but it can take time to access them in your mind. Take the reflective time to allow these important, positive, and optimistic thoughts to bubble up to the surface of your attention. As you reflect, sense what your body is telling you about how and who you should appreciate. Try to connect your appreciation to a positive emotion.
- Share appreciation with colleagues face-to-face. It costs nothing and takes very little time.

PAYOFF
Improve individual and team performance.

CHAPTER TAKEAWAYS

- Embodied intelligence helps you show up with all you are.
- Embodied intelligence takes on new power in the Age of AI, as machines are not made of flesh.

- At the threshold, you know that your body, brain, and environment are closely connected. This means:
  - Prizing rhythms of renewal—mentally, emotionally, physically, and spiritually.
  - Cherishing poetic and kinesthetic intelligence.
- Let go of emotional survival and soullessness. Tune into the physicality of your emotions and into proper selfishness.

8

THE POWER WE HAVE

The body is not a thing we have, but an experience we are.
—PROFESSOR THERESA SILOW

Sometimes we just don't understand the power we have. We can be like Captain Marvel, a fictional character who appeared in comic books published by Marvel Comics. In the film that bears her name, Captain Marvel realizes that she has been fighting with one arm behind her back before she changes her mindset and becomes the most powerful character in the Marvel universe. Similarly, as you embrace embodied intelligence, you can become or remain more powerful than AI. Too often we latch onto AI's superhero or supervillain potential. But humans have vast potential for positive transformational growth. Mental focus, physical freshness, poetic intelligence, and kinesthetic intelligence lie at the heart of such growth.

MENTAL FOCUS

Meet James, normally a successful businessman, but he was in a downward spiral when I started coaching him in the fall of 2019. Here's how James described his predicament.

After arriving at work, I would be in meetings back-to-back from 8:00 a.m. to 6:00 p.m., before I even spent any time with my team. I would then spend from 6:00 to 8:00 p.m. debriefing the team, 8:00 to 9:00 p.m. catching up on emails, and then I could finally do some work. The problem was that my team could have easily taken some of these meetings and debriefed *me*. But I felt too insecure—I wanted to be involved in all the details, so I could answer any question that came my way. In operating like this, I had fallen behind on a different piece of work. A week before the deadline, I had made no progress. To catch up, I worked until 4:00 a.m. every night for a week. I was exhausted. Running on that little sleep, I couldn't think. I was ineffective in meetings. Part of who I am is to inspire others, but I'm not doing it. My heart was racing. I wasn't able to make decisions.

James pushed himself to perform. With little intent to recover, he slid into being overwhelmed. Weeks later, after our first two coaching sessions, James reported this:

I'm spending 50 percent less time in meetings. Team members appreciate the opportunities I now provide. I feel like I can now focus and others tell me I'm more inspiring. I've been able to identify times when colleagues have been visibly irritated by something and intervened to correct the situation.

What enabled James to transform his performance so quickly? Mental energy was a critical factor for him, and we spent considerable time working on this. For the first time in his career, James varied his mental focus between big picture and narrow focus. He made sure to get the most important thing done first every day. He invested ten minutes per day in using a simple time-planning tool, improving his prioritization. We discussed the latest findings about cognitive multitasking: According to hundreds of studies, humans cannot perform multiple cognitive tasks simultaneously without sacrificing quality or speed; and human beings are more effective at performing cognitive tasks sequentially. [1]

I remember when James sighed as he realized how digital distraction assaulted his thinking by forcing unnatural parallel processing on him. "The irony," he said, "is that although many of us multitask to stay tuned into what others think, in fact it stifles that very thinking."

Digital distraction can be dangerous. I recently heard a story about fifteen people sitting on a bus. A hooded man near the back pulled out a gun, strode up to a nearby middle-aged man, and took his wallet. The thief then went to a young woman and took her purse. Everyone was distracted, looking at their cell phones. Finally, one passenger looked up, saw the gun, and decided to fight back. In the first few seconds after this moment, no one else noticed what was happening, because they were still looking at their cell phones. Even the passenger on the bus who initially fought the robber *still had his cell phone in his hand while fighting.* He just couldn't give it up. Several passengers eventually pinned the armed robber to the ground. This story illustrates how technology can blind us to threats to our well-being and make it hard for us to avoid distraction.

Professor Shoshana Zuboff argued that using Facebook does not promote mental well-being.[2] This finding is controversial for some, and any risks are of course not limited to Facebook.[3] The fact remains that some of us can't give up our addictions to machinelike behavior. Technology may be making idiots of us.

By taking the steps above, James not only improved his thinking but also, crucially, connected his thinking to his being. His self-concept improved as he became more centered in his own contribution, and he nurtured others better by delegating tasks to his team and connecting emotionally with clients. James's story therefore also illustrates how closely the four foundational dimensions of embodied intelligence support one another.

PHYSICAL FRESHNESS

The other main factor that fueled James's performance turnaround was a new approach to physical energy, in particular his sleep.[4]

Missing sleep is one of the most hazardous things any human can do.

On October 14, 2012, the "Austrian daredevil" Felix Baumgartner performed his boldest jump yet, some would say the boldest jump by any human being ever. He rode a helium balloon 128,100 feet above Earth into the stratosphere, wearing a 100-pound pressurized flight suit and helmet. In that part of near space, the atmosphere is so thin that blood vaporizes if it's not sufficiently protected. After consulting his forty-point checklist, Baumgartner saluted and stepped off the platform "with as much ease as any of us mere mortals might step off a curb."[5]

Baumgartner was in free fall for four minutes and nineteen seconds and became the first person to break the sound barrier relative to the surface, not using a vehicle. He also broke three world records at the time—for exit altitude, vertical free fall distance without a drogue parachute, and vertical speed without a drogue.[6]

One reporter asked Baumgartner whether he enjoyed the jump.

His response? "Honestly, no."[7]

Enter sleep expert Matthew Walker:

> Struck by the weight of damning scientific evidence, the Guinness Book of World Records has stopped recognizing attempts to break the sleep deprivation world record. Recall that Guinness deems it acceptable for [Baumgartner] to pass through the sound barrier while creating a sonic boom with just his body. But the risks associated with sleep deprivation are considered to be far, far higher. Unacceptably high, in fact, based on the evidence.[8]

Sleep loss crushes not only our performance, but also our neurological, psychiatric, and physiological health.[9] "No facet of the human body is spared the crippling, noxious harm of sleep loss," Walker wrote. He added that we are "socially, organizationally, economically, physically, behaviorally, nutritionally, linguistically, cognitively, and emotionally dependent upon sleep."[10]

I gave James information like this to help him improve the quality and quantity of his sleep. James acted on it. He refashioned his approach to sleep and later reported, "Six weeks on, I haven't always managed seven hours' sleep, but there are only two nights where I've had less than six and a half hours' sleep." This significant improvement brought James physical recovery that helped him bring his emotions to his thinking, which in turn boosted his team's performance.

Remember special forces leader Tony, whom we met in chapter 6? Tony noted that a vital factor in his success on the NATO mission was the quality of his sleep and exercise. Said Tony, "Leading rapid change requires emotional and physical energy as well as clear thought, so I made sure to sleep for more than seven hours a night." Here, Tony applied threshold leadership to help a NATO mission achieve a key goal. How much performance and how many years of life are leaders giving away because they don't prioritize a healthy pulse between physical performance and physical rest? As AI poses more challenges for leaders, sleep will be a crucial differentiator.

Over-separating the mind and body is a cultural pathology common in the West. An example of this pathology is viewing brain function as essentially computerlike. Robert Epstein, the former editor of *Psychology Today*, provided a good explanation of this, saying, "Computers don't play games like humans play games. Computers don't create like humans create. Computers, at their most fundamental level, don't even solve computational problems like humans solve computational problems." Our performance is of a different kind to that of computers, since they don't need to pulse.

The advancement of AI will increase the amount of insight we can glean from data. But our experience of this advancement will be unpleasant if we are exhausted. If pulsing between performance and recovery helps so much, why don't more of us do it? Not, I think, because we don't know what to do. Whether it's quitting an unhealthy habit or addressing that persistently thorny issue at work, I think that we are all simply very smart at finding ways to avoid doing what we know ought to be done. I encourage you to use the resources and inspiration in this chapter to change not only behaviors but also mindsets.

Beyond managing our energy, there is more to embodied intelligence. Next we turn to poetic and kinesthetic intelligences.

POETIC INTELLIGENCE

Like poetry, humans have profound rhythms and powerful emotions. These rhythms and emotions can underpin your performance in the Age of AI, as they connect your brain and the rest of your being physiologically, a connection that machines cannot match. For millennia, humanity has appreciated the power of poetry. Now, it turns out that embodied human poetic intelligence can help you as AI improves.

Take breathing. We breathe not just with our lungs, but with our whole body. Vietnamese monk Thich Nhat Hanh wrote that "Breath is the bridge which connects life to consciousness, which unites your body to your thoughts." In this lies something of what it is to be embodied and what it is to be human.

We also breathe rhythmically. After we inspire, we respire. In, out. In, out. I recently contemplated what is perhaps the best known of Shakespeare's 154 sonnets, "Sonnet 18," which includes the following line:

Shall-**I**
comp-**are**
thee-**to**
a-**sum**
mer's-**day**?

This sonnet uses a line pattern known as iambic pentameter, in which each line includes five pairs of double syllables, with the emphasis on the second syllable in each pair (as in the word a-**bove**). In June 2020, I heard British priest and poet Mark Oakley explain the significance of this sonnet:

As I'm sitting here now, slightly nervous, I can hear my heart going "te **tum**, te **tum**, te **tum**." Poets call that an iamb. A short long. Te **tum**. If I took one breath with you now, I would get five "te **tums**" into my breath . . . an iambic pentameter. And that is the main rhythm of most classic English poetry.[11]

Breath is the heartbeat of poetry. Will AI ever experience what this heartbeat is like? AI can already measure heart rate, analyze research data about respiratory diseases, and help screen for respiratory diseases. But AI doesn't breathe. AI can't compare with our ability to connect our brain and the rest of our being physiologically, to access the power and rhythm of poetry in the inspirational, sometimes breathless way we do. In this sense, accessing our breathing wisely forms part of embodied intelligence.

Next, consider emotion, on which poetry hangs. One of the foremost researchers of emotion is Professor Lisa Feldman Barrett. Feldman Barrett found that bodies play a part in constructing emotion.[12] That is, emotions are physical and embodied. In chapter 1, we saw that machines "have" emotions in a different way from humans, as consciousness probably eludes AI. I do not think that a machine's body, such as its robotic arms, will likely experience emotions in the same sense as human bodies do. Indeed, the fact that AI is not flesh matters more than we think.

Of course, machines and humanity are increasingly integrating. Someone might therefore argue that surely machines, via humans, will be able to have some experience of emotions, in this integrated sense. But that is precisely the point. Threshold leaders will prize the sublime, distinctly human contribution that our embodied experience of emotions brings.

Feldman Barrett also found that culture and upbringing form part of emotion.[13] But AI lacks culture and upbringing in the same way that humans have it. We feel our culture, we know it, in a different way to how an AI can.

Why does the fact that we are flesh make a difference? Because we breathe and our bodies carry emotions. As Mark Oakley observed, "When we fall in love, we are often naturally driven to poetry, whether we or others composed it." We know what it feels like for a summer's day to dawn in our hearts. AIs don't. As a leader, you can inspire others to love their work, love their colleagues, and love their mission by crafting inspiring new stories that have roots in the old. The more you use poetic intelligence, the more distinctly human and effective you will be as a leader. Humanity's most powerful seeds for growth may be in this kind of embodiment, and I think a shift is underway.

KINESTHETIC INTELLIGENCE

In 1910, fifteen-year-old Babe Ruth pitched for the first time in his life. Ruth recalled in his autobiography that "as I took the position, I felt a strange relationship between myself and that pitcher's mound. I felt, somehow, as if I had been born out there and that this was a kind of home for me."[14] Instinctively, Ruth's body moved in exceptionally intelligent ways.

Howard Gardner noted that Babe Ruth "was a prodigy who recognized his 'instrument' immediately on his first exposure to it, before receiving any formal training."[15] There is something irreducibly human about this recognition.

Babe Ruth's intelligence was, among other things, an intelligence of the body, or kinesthetic intelligence. AI drives increasingly impressive kinesthetically abilities. For example, in Hangzhou, China, the manufacturer EP Equipment uses autonomous forklifts that can maneuver themselves in factories and on warehouse floors.[16] The brains of these AI-enabled forklifts adjust to feedback from their bodies, as they learn to navigate and otherwise work with their surroundings. Such adjustments and learning are extremely useful. But the knowledge that resides in this forklift's body is functional and differs fundamentally from the purposeful ways in which humans learn to work with their bodies. In this difference lies opportunity. In the future, embodied intelligence can be a vital part of the caring, loving world you create.

Embodying intelligence does not entail switching off the mind, as if the body is the only way to know. Threshold leaders thoughtfully embrace two key features of kinesthetic intelligence: learning about body movement and using knowledge that resides in your body. We will explore each of these two features in turn.

First, Professor Alison Gopnik of the University of California observed that humans learn to move in ways that differ from machines. She noted that although machines frequently "learn" from the accumulated "wisdom" of previous iterations,[17] this learning fundamentally differs from the type of learning that humans do, which is to learn from "the accumulated wisdom of past generations."[18] Researchers such as Gardner and Gopnik also know that body movement develops in a clearly defined way in children.[19] For example, as Gopnik observes, "Children are active learners; they don't just passively soak up data like AIs do . . . [children] extract information from the world around them through their endless play and exploration."[20] At our best, we learn actively from our earliest days and capitalize on this learning in later life. We embody intelligence in a different way from AI and must cherish this difference, even as AI improves.

Threshold leaders prioritize body movement as a feature of their leadership practice. They know that movement impacts mood, cognition, collaboration, hope, and resilience. As Hillary McBride wrote, "The body is the only way we have to move through life."[21]

Second, our bodies carry knowledge. To illustrate this, we will briefly consider the difficult area of trauma. Psychiatrists recognize that our bodies carry trauma-related knowledge in a different way to how our minds carry knowledge. Along with drug therapy and talking therapy, psychiatrists now view embodied therapy as a valid avenue to help some survivors feel fully alive and move on with their lives. For example, during the last four decades, Dutch psychiatrist Professor Bessel van der Kolk has treated thousands of traumatized children and adults and published more than 150 peer-reviewed articles on memory, neurofeedback, developmental trauma, and other topics.[22]

In his 2014 book, *The Body Keeps the Score: Brain, Mind, and Body in the Transformation of Trauma*, van der Kolk explored important aspects of how our bodies work, drawing on a wide range of disciplines.[23] It turns out that trauma produces actual physiological changes such as "a recalibration of the brain's alarm system [and] an increase in stress hormone activity."[24] Mind and body are more deeply connected than we previously thought. According to van der Kolk, traumatic symptoms originate not just in our minds but in our entire body's response to the original trauma. These are distinctly human findings.

Knowledge resides not just in our minds but in our bodies.

Van der Kolk's work intersects with findings ancient and modern that also support this integral, less dualistic, perspective. Thousands of years ago, the word *nephesh,* in biblical Hebrew, carried a variety of meanings. In a human context, it encompasses the idea of the whole person, including the body. More recently, as we have seen, Feldman Barrett contended that we always look at reality through the lenses of culture and upbringing, which affect our physically embodied emotions. And, of course, culture and upbringing are common sites of trauma. The body carries socially constructed knowledge, memory, and experience in a way that a neural network cannot.

Your body is central to how you access knowledge. Those who discount the deep connection of their thinking with the rest of their being lose a kaleidoscope of inputs and ignore millennia of wisdom.

Every body holds knowledge. Not just everybody. *Every body.*

KINESTHETIC COLLABORATION

Unthinkingly led, AI will magnify our terrors. Well led, AI will enhance human flourishing. But how far could AI get with kinesthetic intelligence in the future? Not very far on its own, in my opinion.

You might argue that AI is already very far ahead. Consider this argument by analogy: DeepMind's AlphaGo Zero, introduced in 2017 in the science journal *Nature*, plays the game Go at a superior level to human players and received no training beyond receiving the rules.[25] And in 2019, DeepMind released MuZero, a more generalized program that achieves industry-leading performance in fifty-seven Atari games. MuZero also matches AlphaGo Zero's performance at chess, shogi, and Go, *all without being told the rules of the games in advance.*[26] In other words, after coming into being, MuZero taught itself the rules in an interactive way and then won. What is the difference here between AI and humans? My eldest daughter once declared that she would teach herself to swim, refusing my offers of help. (In a rare showing of fatherly wisdom, I let her do what she wanted, and it worked.) As robots become more dexterous, won't embodied AIs soon excel kinesthetically without needing to have been trained?

This is a serious and credible argument, to a point. The Santa Fe Institute's Embodied Intelligence project seeks "a theoretical framework that will guide the creation of artificial agents that adjust their neural networks (brains) to feedback from their bodies and surroundings—in essence to learn how to navigate their surroundings."[27] This is a worthy and necessarily limited aim. Responding to physical or even emotional feedback from the environment is a far cry from carrying memory in an artificial body, not a neural network. Nothing in the custom tensor processing units and training methods of AlphaGo Zero and other AIs indicates the imminent arrival of artificial kinesthetic intelligence that carries knowledge in a way that closely resembles how human bodies carry such knowledge. The more life-giving route will be to work with AI to capitalize on *human* kinesthetic intelligence.

There's a deeper point. Humans use Go to learn about other dimensions of life, not just to win. These lessons include the importance of flexibility, patience, and responding well to seemingly grand reversals. Our embodied intelligence opens vistas of meaning that are unavailable to AI, as we use physical life to usher in a variety of purposes.

Poetry and movement live in us, rhythmically, viscerally, in a way they do not with AI. We experience body language, moving in three-dimensional space, the sound of birds, the smell of trees. It is not obvious that we can reduce that to 1s and 0s. Rather, evolve your leadership at the threshold by integrating your thinking with your embodied being. You are more than a brain on legs.

THRESHOLD RESOURCES

RESOURCES FOR ANY LEADER

Resource 13. Peace with Your Breath
PURPOSE
Embody peace when you are stressed.

PROCESS
Follow this poetry-fueled breathing exercise, which is slightly adapted with permission from an exercise written by Brian Draper.[28]
    Pause for a few moments, wherever you are, to sit and engage with the following simple exercise:

- Relax your body.
- Slow your breathing down, and deepen it a little.
- Bring all of your attention to the in-breath, as you inhale, and then the out-breath, as you exhale.
- Smile.
- Be still.
- Close your eyes.
- And keep going, for at least a minute or so. But preferably longer.
- (Thank you.)

What did you notice?

Brian's friend Ciaran wrote to him recently about breathing. He'd seen the inspiring film *Breathe*,[29] about the pioneering polio survivor Robin Cavendish. Cavendish was kept alive in the 1950s by an iron lung but was determined not to be shackled for life to the machine.

"Did you realize that a sixty-five-year-old like myself will have taken over five hundred million breaths so far?" Ciaran asked. "I wonder how many we have been aware of?"

It's a good question! It's so helpful to pay attention to our breathing. It calms us and truly helps us become more intelligent, for the good of others and ourselves.

After all, we don't want to just cope with this wonderful but sometimes stressful time, do we? We surely want to flourish! As Robin Cavendish said in the film, as he fought for a revolutionary mobile form of respiratory machine:

"I don't just want to survive, I want to *live*."

And so say all of us.

Bring your attention back to your breath.

As you breathe, simply and compassionately observe how you are doing right now.

As you sit, keep your back straight to help keep you alert.

Take some more deep breaths in through the nose and out through the mouth. This maintains the healthy balance of oxygen and carbon dioxide that your body needs to function well.[30]

Why not carry some of the goodness from our breathing exercise into your activity this week, and bring your attention back to your breathing as often as possible?

One excellent way to do this is to read poetry while sitting under a tree or by a natural water source. As poet Michael Longley wrote, "If prose is a river, poetry is a fountain." Many leaders rush. Sitting, simply sitting, can escort us into sublime, refreshing places.

As you sit and read, breathe deeply. Allow what you read to infuse your mind and body. You might try Tess Ward's *The Celtic Wheel of the Year*, which begins with these words:

Living Presence, I come here this morning looking up at the sky,
longing to see you and have my soul earthed in security.
Embody yourself in all the life round me this day.

Alternatively, try one of the following:

- Read this book: Olav H. Hauge, *Don't Give Me the Whole Truth*.
- Listen to Eric Whitacre's meditative song "Lux Aurumque."
- Read Mark Oakley's excellent poetry anthology, *The Splash of Words*.
- If it is your practice, reflect on the notion of God breathing through you. Set an intention to embody something more of God's love and peace, at the following times:
  - when you're sitting in a traffic jam or at a red light;
  - when you're at the shops;
  - when you're at the back of the queue and feeling impatient;
  - when you're chatting to someone you've bumped into;
  - when you're pausing to savor a coffee;
  - when you're taking time out for silence and stillness;
  - when you're walking in the woods;
  - when you're at the doctor's office or having chemo;
  - when you're staring at the moon;
  - when you're eating your supper;
  - when you're heading off to bed . . .

PAYOFF

- Become a more inspired leader.
- Connect old and new in resonant and inspirational ways.

Resource 14. Increase Your Sleep Quality

PURPOSE

- Improve the quality of your sleep.
- Enhance your leadership contribution, even as machines do more.

PROCESS

This practice takes the form of a few tips. It does not include a thorough process of diagnosis and resolution, for example for those suffering from chronic sleep problems.

Many leaders underestimate the importance of even small improvements in the quality of their sleep. The research underpinning the benefits of sleep is vast, wide-ranging, and impressive.

Here are some useful tips:

- **Set a sleeping alarm.** Set an alarm an hour before you want to go to sleep. During that hour, do calming things like have a shower, drink a noncaffeinated hot drink, read a relaxing book, or have a gentle conversation. (Nothing involving blue light.)
- **Consume less caffeine and alcohol.** (Or have it earlier in the day; maybe not too early in the case of alcohol!) "Alcohol is one of the most powerful suppressors of REM sleep that we know of . . . one function of REM sleep is to aid in memory integration and association."[31] Caffeine late in the day is little better.[32]
- **Cool your room.** Set the temperature of your sleeping room on the cool side. The Sleep Council recommends 60–65°F (16–18°C) and the Sleep Foundation recommends around 65°F.
- **Keep an "ideas pad" next to your bed.** Keep a pad of paper and a pen by your bed. If you wake during the night beset with worries or churning ideas, make a short note of the idea or worry. As David Allen said, "Your mind is for having ideas, not holding them." By making a note in this way, you remove thoughts from your working memory in the night.

PAYOFF
- Let go of tiredness.
- Become less overwhelmed.

Resource 15. Knowledge That Resides in Walking
PURPOSE
Lead by walking.

PROCESS
Like water, leadership goes stagnant if it doesn't move. One of the best ways to move is to walk. Here are three ways you can lead by walking.

A. WALK THE FACTORY IN PEACE

Several of my clients work in factories or lead business units comprised of factories. When you visit these sites, take unrushed time to walk around.

It's so easy to walk through life preoccupied. Instead, practice walking not to arrive. Drawing on the wisdom of Thich Nhat Hanh, Brian Draper presents this alternative:[33]

> Walk, instead, in such a way that "Peace is every step." And by learning to embody this physically, it can help us mentally, emotionally, and spiritually to be a little more present and at peace.
>
> [Thich Nhat Hanh] says, be aware of the contact between your feet and the ground, and "Walk as if you are kissing the Earth with your feet. Walk in that spirit."

When at the site, take fifteen minutes to walk, not to arrive. As you walk, tune into what your body is telling you. Reflect on the following questions:

- Where in the factory do I feel heavier and where do I feel a lightness in my step? Dwell longer in these places.
- As a leader, what does it mean for me to walk in peace?
- If my organization is a body, what is that body saying to me?
- If the earth is a body, what is that body saying to me?

By walking in peace and in an unrushed way, you can role-model transformation. You can become the very answer that others might otherwise struggle to express adequately.

B. WALK LIKE SOMEONE ELSE

Walking like someone else is not about becoming or mocking someone else. It's about raising your awareness of your own body by experiencing what it's like to walk differently. There are two ways you might try this:

- Agree with a trusted team of colleagues that you will set aside fifteen minutes for the following exercise during an off-site or longer team meeting:
  - Stand in a large circle, facing the person to your right.
  - Pick someone to be the leader.
  - When the leader wants, start walking around the room or even the entire venue (if sufficiently brave).
  - Everyone in the circle copies the leader's manner of walking. Try mimicking gait, posture, and other aspects of how the person walks.
  - After a while, switch leaders.
  - Return to where you started and reflect on what you learned. Encourage each person to engage with what bodily sensations they experienced during the exercise and to consider what they learn from that.
- When you're walking down the street, try adjusting your walking pace to the pace of someone in front of you (not for too long, or else that person might get suspicious). Increase your awareness of how you are walking as you follow.

Such exercises often surface deep reflective insight because our bodies carry so much knowledge. My friend Jeroen Drontmann once said: "These reflections fly away in the wind of rumor and data. Before you know it, they are gone." Take a moment to write down your reflections after each exercise.

C. WALK THE EXTRA, LOVING MILE

Walk to someone else's work area. That space may be the person's open plan area or lab. You may want to check that you have permission first. As you step into his or her space, know that you are a powerful presence. You bring your whole self.

Here are some ways in which you could lovingly serve that person:

- Give a physical gift that symbolizes a positive quality you see in him or her.

- Kneel and ask forgiveness, if you have wronged the person in some way.
- Help that person stuff envelopes.
- Sing a song or read a poem that celebrates something the person has done.

In the above ways, you go the extra mile beyond email to embody transforming love.

Payoff
Become a moving leader.

Resource 16. Do a Digital Detox
Purpose
Unplug from digital devices, together with your team.

Process
Much has been written about individual detox, so here are my tips for enjoying a digital detox with a team:

1. **Expect it to hurt.** Don't go into it half-heartedly. It's called "detox" for a reason. Discuss as a team what your barriers to detox success are likely to be.
2. **Break the challenge down.** Start by designating a shared thirty-minute period each day to switch off your devices and Wi-Fi. Or commit to reduce your phone usage by over one hour per day, by using the Moment app. Or commit to each other to keep your phones downstairs at night and not in your bedrooms. After all, a 2017 University of Texas study showed that your smartphone reduces your cognitive capacity if it's near you, even if it's off.[34]
3. **Turn a negative into a positive.** Shift your mindset from "I'm not going to check my phone" or "I'll avoid Netflix" to "I will focus fully on my team members and on my task priorities." Hold one another accountable for being present.
4. **Bring structure to your detox.** Discuss as a team how, during detox, you might attend to the following ten areas: the

environment; diversity; exercise; leisure and hobbies; your partner, spouse, or significant other; family; friends; your true self; work; and personal growth. In each of these ten areas, you can have a very positive impact.

5. **Kick-start your detox** by going on a seasonal retreat together with your team.

John Shedd once wrote, "Ships in harbor are safe, but that's not what ships are built for." Detox can feel risky, but when you prioritize coherence over fragmentation, depth over shallowness, freedom over dependence, the results are worth it.

PAYOFF

- Reassert your independence and put your devices in their place.
- Model healthy technology habits for a world that is sometimes blinded by technology.

Resource 17. Body Communication

PURPOSE

Embody your communications.

PROCESS

Communication lies at the heart of leadership. Here are three ways you can embody communication and inspire your team to do the same.

A. COMMUNICATE BY BEING SENSITIVE TO EMBODIED EMOTION
   When Oprah Winfrey asked Amanda Gorman about how she wrote "The Hill We Climb," Gorman replied that she would sometimes spend hours on a single word or phrase. You may not choose to spend that amount of time on a single word or phrase. But when preparing speeches or updates, do invest the time to tap into sources of deeply held meaning. Where possible, get lyrical with vision statements and use poetic metaphors to inspire, as this will increase your embodied connection with your message.
   Communication also includes listening. When listening to colleagues, listen for emotion, not only for the logical content of what

they say. Remember: Emotion is bodily. Here are some considerations that may help you sense another person's emotion when
they speak:

- In general, how are they sitting or standing? Are they taking up
  a lot of space? Are they curled up in ball? Are they edging away
  to the corner of the room? Different ways of physically showing
  up can reveal different emotional states.
- What physical manifestations of emotion are they showing at
  different times, such as blushing or shaking?
- What is their pace, volume, pitch, and breathing like as they
  speak?
- What emotion words are they using, or other words that feel
  "heated" to you in some way?
- What are they not saying? How are they not saying it?

B. COMMUNICATE BY HOW YOU HOLD YOURSELF.
Threshold leaders are good at checking into where they are in their
body at any given time.

Increase your connection to your body via physical activities such
as Pilates, aikido, or posture work. As Paula Mariani wrote, "Our body
continually reminds our truth as a fact we can't deny."[35] Bessel van der
Kolk refers to "learning to inhabit your body."[36]

C. COMMUNICATE BY GIVING FEEDBACK WITH BODY AWARENESS
Cultivate awareness of what is going on in your body when you give
or receive feedback. Maybe you start sweating when you have to
deliver a difficult message. Or maybe your chest tightens every time
you realize someone is about to give you developmental feedback or
even praise.

Increase your ability to recognize present physical sensations by
practicing feedback situations in advance with a trusted colleague.
Doing a role-play of this kind can take as little as ten minutes. Discuss
with this colleague how you felt bodily at various points in the role-
play. By raising your awareness of such feelings and sensations, you
strengthen your leadership presence.

Payoff

Inspire those around you.

CHAPTER TAKEAWAYS

- Not only are humans not machines, we are not like machines. Sleep is an underused lever to capitalize on this insight.
- Used rightly, your mental rhythms are deeply powerful.
- Threshold leaders also embody intelligence through:
  - speaking and writing in evocative, inspirational ways,
  - breathing well, and
  - capitalizing on knowledge that resides in their bodies.
- The possibility of techno-humanism catapults embodied leadership even more firmly to the top of the agenda.

9

DRY WOOD MEETS FIRE

*When people are connected to their bodies and emotions, they
are almost impossible to control.*

—JEMIMAH MCALPINE

In 2018, I spoke with the Financial Services Group (FSG), part of the Chinese technology company Baidu. The FSG was harnessing AI to simulate
people's behavior and develop financial products. With a valuation of $3
billion and an operating income of $300 million in 2017, the FSG was a
significant player in its markets. In this chapter, we will learn how FSG
nurtured a culture that welcomed organizational embodied intelligence.
I define *organizational* embodied intelligence as the capacity to embed
individual embodied intelligence into processes, systems, infrastructures,
and other aspects of culture.

By the summer of 2018, it was clear to the management team that this
would be a challenging year for financial services. The sector had become
extremely risk averse following the 1998 Asian financial crisis, the 2008
sub-prime mortgage crisis that had spread to China, and now the 2018
peer-to-peer lending crisis in China. Among other initiatives, the FSG
sought to use AI to address financial exclusion, which was an increasingly
pressing social challenge in China.

Everyone I spoke with at FSG shared a beautiful generosity. They love
to give of their time and creativity to colleagues, which will in turn serve

wider society. "Financial Services meeting AI is like dry wood meeting fire," an FSG leader told me, speaking on condition of anonymity, "because of the possibility of widespread financial exclusion driven by inexplicable algorithms. We were facing this technology challenge at the same time as drawing together and inspiring disparate parts of the business to create a team of eight hundred people, and then growing it quickly to eighteen hundred people."

What was the exclusion challenge that the FSG faced? As of late 2018, of the 900 million economically active people in China, more than 550 million lacked a credit record with the central bank of China. A further 100 million people had credit records rated as poor, but this group of more than 550 million were typically in an even worse position in relation to getting loans, as they lacked a credit record at all. Among this group were many who had left high school without graduating and wanted to pay for vocational courses such as hospitality, hairdressing, or coding. But lacking a credit record, they could not access the money they needed.

Eventually, the FSG established personalized lending, engaged its teams, and increased Baidu revenues. Let's analyze how it accomplished this.

On the surface, the FSG's approach mirrored that of hundreds of technology firms, as it followed a three-step process to create financial inclusion for those lacking a credit record. First, leaders sought to understand who the excluded people were and how they typically behaved by using data in Baidu's ecosystem. "We have a sustaining partnership with Baidu Group's Big Data Division, and this is a win-win situation: FSG needed data to build AI models, which later have proven to be effective in recognizing underserved borrowers, acquiring customers, and assessing risks; while the Big Data Division is commercializing its data assets through industry-specific solutions."

Many of the Baidu apps have strong market share, such as Baidu Map, Baidu App Stores, and Baidu Read, and hence data from these apps offer unique and valuable input to understand user behaviors and generate potential customers' profiles. In this way, the FSG understood what type of reading applicants usually did, their learning habits, and what apps they downloaded and installed.

Second, FSG developed proxies that it could use to justify credit.

Third, it split applicants into different, small segments, offered micro loans, and observed people's behavior in a series of A/B tests. Knowing how and when different groups of people paid back their loans, FSG assigned credit risk scores that it used to determine how much money to lend, for how long, and at what rate of interest. It also used this information to match different products to different people intelligently and to predict likely outcomes. Narrow AI exceled in this, given the multiple variables and vast amounts of data involved.

More deeply, the following three markers of an embodied culture also turned out to be vital to the FSG's success:

1. Poetic and other existential input
2. Embodied energy rhythms
3. Knowledge that resides in bodies

Collectively, these markers encompass the six dimensions of the foundational and advanced levels of embodied intelligence described in chapters 7 and 8. We will now consider each of the above three markers in turn, before looking at cultural and ethical reasons embodied organizational intelligence matters for teams and organizations.

POETIC AND OTHER EXISTENTIAL INPUT

The FSG management committee met weekly. Committee members included the FSG CEO, three Baidu vice presidents, and executives of major functions covering technology, strategy, finance, and human resources. This was a diverse group, and top leadership integration mattered. The committee used its weekly meetings highly creatively, raising awareness of its culture in relation to the rest of Baidu and challenging this culture where appropriate.

The FSG's management committee members devoted half of each weekly meeting to discerning who they really were collectively. The management committee weekly meeting is a well-enforced and continuous mechanism, and one critical topic that management members attended to, in addition to who they really were, was how to innovate and create value. For example, during these afternoons, members drew on visual and other resonances of Mandarin and used visualization to identify their shared purpose and learn about one another's motivations. They then used those insights to drive organizational change in an aligned, centered way. They also used a wider, regular, quarterly leadership team meeting to deepen their shared sense of values and core assumptions that drove the organization. Through these innovative embodied processes, teams at Baidu and FSG intentionally sought different perspectives and uncovered a shared sense of true self, which set them up for success.

EMBODIED ENERGY RHYTHMS

During the FSG's fast-growth phase, members of the senior team discovered that emotions drive performance and that thinking and feeling are bodily things, not just cerebral things. As a result, many middle-level business leaders experimented with generating good mental and emotional energy rhythms.

For example, early in the FSG's formation, it faced the issue of whether to diversify into the potentially profitable sector of insurance. Advocates saw that AI could drive lower property and personal insurance costs, for example, via fast development of drugs that fight diseases. This provided a systemic reason to diversify into insurance. Against this, however, were the performance, security, and reputational risks associated with getting too diversified too early. Balancing such benefits and risks is complex work.

The team worked through this by varying its focus between, on the one hand, narrow analysis of the benefits and risks and, on the other hand, big-picture thinking about what was best for the FSG, Baidu, and its customers. They also practiced appreciation, empathy, and affirmations of trust. In other words, FSG prioritized mental and emotional rhythms that drove physical recovery, finer thinking, and motivation. These were embodied moves. As one early FSG-er and also insurance-business-builder explained, "Allowing people time to recover emotionally and mentally was a competitive advantage that helped us win."

KNOWLEDGE THAT RESIDES IN BODIES

Alongside the work above, the FSG leaders faced complex questions such as the following: What if the present mix of human involvement in financial services—80 percent working on efficiency and 20 percent working on effectiveness—were to change drastically? To address this question, the management committee members paid close attention to the physical spaces in which they met. Some ad hoc lunch meetings would be followed by walking meetings. Some executives began to lead their team to view body movement as a crucial plank in thinking well. They understood that even in a context of highly advanced AI, there is no substitute for the finest thinking of embodied human team members.

An experienced FSG strategist told me that, as leaders engaged their bodies in these ways, they became more effective at examining assumptions that underlie complex questions, and they had a more positive attitude toward internal debate and challenging questions. "Nurturing such an open mindset was central to the value that our strategy department created," explained this strategist. In large part, practices such as these drove the success of the team.

As the FSG management committee walked and embraced creative, embodied rhythms, they opened the way to draw on this deeper knowledge. They recognized that the FSG could only operate effectively if it had a culture of prizing a range of intelligences. What if you were to explore a similar journey with your team?

My discussions with people at FSG inspired me to reflect further on why embodied intelligence matters for organizations and societies. As I considered embodied organizations in the United States, Europe, Africa, and Asia, I realized that teams and organizations with high embodied intelligence share three features:

- Cultural vitality
- Ethical thoughtfulness
- Uncontrollability (sounds a little out there, but I explain below)

Let's consider these factors in turn.

CULTURAL VITALITY

If we do not embody intelligence organizationally, we abandon our colleagues and societies at the time they need us most. This is at heart a cultural point. As Hillary McBride noted, embodiment is "a way to heal the mind-body divide we experience within ourselves and, more systemically, within Western cultures."

A few months after the first management committee meetings, the shared reflection of FSG colleagues was that they had developed an embodied culture of leadership at scale. FSG's cultural initiative motivated others and inspired them to seek different perspectives. These two leadership behaviors, underpinned by creative mindsets, characterize high-performing organizations.[1] They are also embodied behaviors, as motivation and perspective arise not just from our brains, but also from the rest of our body. In prioritizing these behaviors, FSG teams intuitively realized that flesh matters, and they created a culture of embodied leadership. In FSG's case, embodied intelligence led to a market-leading AI transformation.

Although some embodied mindsets or behaviors are only subtly different from traditional ones, the difference in effect is huge. Robotic action purely rooted in logic is just not the same as action that arises from a broader multiplicity of intelligences. Leading purely cerebrally in a moment of AI challenge is—as the saying goes—like rearranging deck chairs on the *Titanic*.[2]

Cultural vitalization occurs most effectively throughout an organization when it occurs at scale. To accomplish this, leaders at the threshold do not just create a culture of embodied leadership development; they also pay heed to performance management, talent acquisition, talent development, leadership mobility, succession planning, and organizational development. My colleagues and I regularly work with leaders to optimize such broader mechanisms, to transform culture, and to deliver leadership at scale in an embodied way. Done well, such work is at the heart of the threshold.

Transforming an organization using this kind of blueprint can be messy. But recall Bourgealt's warning that disembodied thinking creates "a superficially tidy universe [where] everything is in a box." In your organization, culture will advance from superficial to real the more you recognize that your colleagues and customers are embodied agents. As you lead, will you prize embodied emotions? Will you vitalize culture by cherishing the knowledge that resides in your own and others' bodies?

ETHICAL THOUGHTFULNESS

Organizations that embody intelligence can address ethical questions in a more sophisticated way. For example, one of the biggest ethical debates the FSG faced was about their use of data. On the one hand, customers stood to get cheaper, more personalized products the more Baidu held and used their data. On the other hand, many feared a loss of privacy. "Search engines probably know more about you than your wife or husband do," one big data veteran joked. "When you search for something online, you will tell your search engine but you will not necessarily tell your spouse." How would or should the FSG balance these considerations?

This was not an easy question for FSG. Indeed, it is a challenging question for all internet giants around the world who seek to operate financial services businesses. A customer's search history and online behavior can indicate his or her ability to repay loans, and this is a crucial factor in deciding whether to authorize borrowing. For example, it is not difficult for technology platforms to know how many apps users have installed and how active they are in those apps in a given period, how many places they have traveled to and at what time, and whether they click and spend time on gambling websites. If they chose to, Baidu could use this data to make loan decisions based on what some people may have counted as secrets.

It became very clear to FSG and Baidu Big Data that this misuse or even abuse of data was not only illegal but would also eventually hurt business, as customers would become less engaged. The more they knew or suspected about data harvesting and alleged increased political interference in social media during elections, the more passive users they were likely to become.

In this context, the FSG nurtured an embodied culture by using its afternoon meetings to check in with how each person on the team really felt (emotionally, physically) about this ethical debate, recognizing the wider context in which the FSG operated. Remember, emotions are physical, embodied things. Threshold leaders inspire their teams to listen to what emerges from their inner sources of knowing. In the FSG's case, leaders also sought to learn from customers rather than try to solve this ethical debate on their own.

In relation to customers, the FSG team members saw potentially misaligned incentives. If they had wanted to, they could have tried to maximize shareholder value by pushing loan products with the most profitable terms. But such products may not have matched customers' credit ratings, and some customers would have lacked appreciation of the risks they were taking in signing up. Therefore, the FSG team sought to use knowledge that resided in their customers, not just in their own team. They did this by conducting customer research including focus groups, where they identified key assumptions that some customers held, assumptions such as "Baidu will not self-regulate well."

In all this, let's remember that data privacy is a very sensitive issue. FSG key leaders navigated this by committing to balance three things: 1) operate strictly within the boundaries of laws and regulations, 2) respect customers, and 3) create value out of data through viable technology and governance solutions. For example, an FSG-er told me that Baidu's Big Data Division never transfers individual data to FSG, but only processed data such as labels. In addition, major data transfers need collective approvals from business leaders, data leaders, and even the data management committee at the Baidu Group level.

This threefold balance is difficult to sustain. But the embodied way in which many FSG-ers led, described above, helped FSG keep things fairly well in balance, which fueled its success. Embodied threshold leaders access an organizational competitive advantage.

Throughout my career, I have observed some team leaders in large organizations disconnecting from their wider context and starting to look down on their customers, competitors, or other stakeholders as somehow less than human. This disconnection is an example of superficial, disembodied thinking. It hampered these executives' ability to address ethical questions well and hampered the value of their contribution, as they tapped into their intuition less and no longer brought their full selves to the questions.

Against this, the FSG organizational "body" sought out knowledge that resided in external "bodies," in this case, customers. By moving physically into the same spaces as their customers, they made the significant move of valuing equality between themselves (app architects) and those without a credit record (potential customers). They avoided the mistakes of demon-strating "the ignorance of contempt prior to investigation," to use a phrase sometimes attributed to Herbert Spencer, and of viewing users as what Alan Jacobs called the Repugnant Cultural Other.[3] They also decided that the potential costs of learning their customers' moral dialect were worth the investment, so they made an emotional connection with them.

Baidu continues to grapple with thorny ethical issues. No technology firm will solve them alone. Maybe you value embodied intelligence but work in an organization that doesn't get it. For example, some organizations tend to reject workers who talk about painful events, which then causes these workers to withdraw. In such a case, you might seek out champions for safe spaces where workers can express pain that they may be carrying mentally, emotionally, or physically.

Author Robyn Henderson-Espinoza views the body as a social reality, not just material "flesh and bone, consciousness and affect." Writing from deep personal experience of embodiment, Henderson-Espinoza observes that the body is a borderland, "a paradox of being and becoming." These insights thrill me, as they illustrate the complexity and possibility inherent in our embodied selves. This impulse of paradox creates a trajectory toward cultural vitality and ethical connection . . . as well as uncontrollability.[4]

UNCONTROLLABILITY

Teams and organizations who lack embodied intelligence are easier to control. If you think that controlling others is a good thing, I'd point you to my historical sketch of organizational development (see "Meet the Threshold" in this book). Models of leadership characterized by control are no longer fit-for-purpose. In addition to this, if our societies normalize the view that the body is just a controllable object in which we reside, little moral force prevents our subjugation.

In the 1999 film *The Matrix*, directed by Lana and Lilly Wachowski, robots prevailed in a nuclear war and rounded up humans to use them as power sources. Twenty years after I first saw the film, the image of rows of subconscious humans wired into vats of orange sticky liquid remains seared on my memory. What is really wrong with using humans in this way, if our brainpower is all that matters?[5]

As embodiment researcher Jemimah McAlpine put it,

> Once you stop trusting your emotions and your body, you need an outside source to tell you what to do, what's right. This happened through history, for example, where the church suppressed dance in worship—it was too equalizing, too empowering. They quashed a sense that people are connected to their bodies. When people are connected to their bodies and emotions, they are almost impossible to control. They have a deep sense of what's right and what feels off.[6]

I do not pretend that embodied intelligence will always be easy to promote or apply. But in a world of increasing polarization and demagoguery, it may be our most important intelligence. The FSG embodied intelligence to some good degree, despite wider constraints. Our future society will be healthier the more we all embody intelligence. Otherwise, leaders may enable a world in which AI disembodies us and devastates society.

Those who are almost impossible to control fill healthy organizations. They are the creatives who inspire action, the threshold leaders. Such leaders embody intelligence and are best positioned to inspire a beneficial AI world.

CRYSTALLIZING EMBODIED BREAKTHROUGHS

I recently encountered a thought-provoking story in Greg Baker's book *The Energy Equation: Unlocking the Hidden Power of Energy in Business*. This story inspired me to think about embodied intelligence through the lens of crystallization. This is a useful lens with which to draw together the threads of this third threshold pathway.

In his book, Baker describes research conducted by the renowned Japanese scientist Masaru Emoto.

> Dr. Emoto [exposed] water from pure springs to both positive and negative thoughts, words and energy. For example, he wrapped pieces of paper with words on them around bottles of water and, following a period of exposure to the words, froze the water. Phrases like "Love and Gratitude" and "Thank You" had a positive effect on the water that produced beautiful, clearly formed crystals.

Conversely, water samples exposed to words like "You Fool!" and "You make me sick. I will kill you" formed no crystals at all. I was absolutely dumbfounded by these results. Although they made intuitive sense to me, it was striking to see the intangible effects of energy made tangible in a way that could not be denied or ignored. Emoto took his research into the realm of human interaction when he adopted a Japanese elementary school as his next testbed. He brought four samples of water from the same spring and instructed the children to treat each bottle differently. To the first bottle they were to say, "You're cute." To the second they were to say, "You're beautiful." To the third they were to say, "You fool." As for the fourth, they were told to completely ignore it. The children complied and the samples were then frozen. The first two, You're Cute and You're Beautiful, produced amazingly beautiful crystals. The third, You Fool, produced distorted crystals, and the fourth, which was ignored, produced the most distorted crystals of all.[7]

This striking story posits a strong link between the physical world and the world of ideas. "It was clear that . . . positive and negative energy was effecting [sic] the water on a molecular level," Baker noted. "Emoto said it was all about vibration, a form of energy. Considering the fact that approximately 60–70 percent of the human body is water, he extrapolated that the energy around us, and within us, effects [sic] our health and well-being as well."

Whatever you think about the details of the above story, it does seem that our thinking interweaves powerfully with the physical world. Every week I see the link between embodied energy and intelligence playing out among leaders. I am also struck by the role of crystals in the story above. What a beautiful metaphor for our sparkling, emergent future!

How will threshold leaders crystallize organizational, embodied intelligence? The research of Dr. Otto Scharmer provides a useful way into this question. Although not without his critics, Scharmer has established himself as an authoritative voice on leading from the future.[8] In his book *Theory U: Leading from the Future as It Emerges*, Scharmer invites readers to imagine an abyss: "One part of our self is on the left side—in the current reality, looking into the abyss. Another part of our self, our emerging Self, is already operating on the other side—the side that connects to the future that wants to emerge."[9]

Scharmer seeks to connect thinking and being in an embodied way.[10] For Scharmer, crystallizing means "clarifying vision and intention from our highest future possibility . . . [which] happens from a deeper place of knowing and self [than normal visioning]."[11] Three of Scharmer's principles for crystallizing what I view as embodied intention are:

1. Let come: "Listening to what emerges from one's inner sources of knowing," where leaders are hyperaware of their bodily sensations.

2. Grand will: "Acting as an instrument of the emerging future and bringing it into reality as it desires." For example, Scharmer notes that "when the conversation drops from one level to another, [you] can feel it in your whole body."

3. Venues for waking up: Setting the right infrastructure for crystallizing intention. For example, Scharmer advocates "mov[ing] yourself outside your organizational boundaries," and he connects twelve management functions with the process of breathing.[12]

These principles and traits are distinctly human. Threshold leaders prize and prioritize them. The resulting intention crackles with power. It's the difference between your team feeling the possibility of a beneficial future and your team knowing in their whole being that it's more than just a possibility; they can't not do it.

Although not using the term *crystallized intention*, the FSG management team members cherished the principles above. They used their afternoon leadership meetings to listen deeply and establish a strong embodied foundation for their organization's AI vision. In doing so, they sensed and acted on a larger will than just their own, a grand will that also included FSG's, Baidu's, and their customers' wills. Finally, the FSG was also prepared to challenge its own embodied thinking, using "venues for waking up" that invited multiple perspectives.

FSG is not all about success stories. Like many AI-driven business innovators, it has seen a high turnover of management team members. AI transformation affects not only traditional business performance but also organizational health, including culture. Therefore, leading AI change is especially tough. Embodied threshold leaders are needed, as they inspire deep trust among their senior team, even when this team changes and evolves more than in other organizations.

In our shared future, threshold leaders will be needed more than ever, as they will be the ones who are embodied enough to inspire us to see AI as a source of hope, inspiration, and consciousness, rather than as a source of threat. More than this, embodied threshold leaders will ground us in what matters most, guiding us to reclaim AI as necessary and refashioning our view of leadership from role to way of being.

Together, the first three pathways of stillness, independent thinking, and embodiment catapult leaders from merely directive, logical intelligence to generative, multiple intelligences. In this sense, the first three pathways are dynamic, as they progress beyond a merely logical approach. But in another sense, the first three pathways are static, as they do not include what psychologists call a model of increasing human development. As Superintelligence looms, this kind of human evolution will be critical and is therefore the subject of Path IV.

THRESHOLD RESOURCES

RESOURCES FOR LEADERS IN LARGE ORGANIZATIONS

Resource 18. Embodied Transformation Visualization
PURPOSE
Develop an outline embodied leadership transformation story relevant for the Age of AI.

PROCESS
This is a team or departmental reflection exercise that takes between three hours and two days to complete. The exercise includes the following five steps:

1. Visualization
2. Composition
3. Recording
4. Performance
5. Direction-setting

This exercise works best if you find an inspiring place to do it, preferably outdoors in a beautiful, spacious place where people are likely to feel connected to nature.

Before starting, ensure each person has a good quality notepad and pen, or a digital equivalent that does not have a cellular or Wi-Fi connection.

Encourage each person to sit in a comfortable position. Use the following script as you find it helpful. Build in time for each part of the reflection, according to your schedule.

STAGE 1: Visualization

Visualization is important, because transformation starts in our hearts, minds, and bodies, and only then goes outward.

Close your eyes or rest your gaze softly in the middle distance. Direct your attention outward, noticing the sounds and smells around you. Smile. Gradually turn your attention inward. Notice first if you have any stress or physical tension in your body. Try to relax your body from your head to your toes. Take a few slower, deeper breaths. Allow any tension to leave your body as you breathe.

Hold your attention on what is and what matters, rather than on how you want things to be, before moving on.

Now move on to imagine a future, three to five years ahead, in which you feel delighted, inspired, fulfilled, joyful, and at peace. Begin to picture a future in which AI is more advanced than today and in which you feel satisfied. Use the power of your imagination to create a picture of your ideal self. In your ideal future, where are you, who are you with, what are you doing? What are people saying to you or about you? In your ideal future, what sort of AI leader are you?

Take another deep breath.

Now imagine that you have fallen in love with that future. Head over heels besotted. Enjoy imagining this. What emotions arise for you? Be playful as you reflect. Smile and relax. Open your eyes.

STAGE 2: Composition
Compose a poem or a song of love to this lover—your desired future or vision. It could include gratitude, humor, soaring eternal themes, and whatever else you feel moved to include. As you compose, you may find the following guidance helpful:

- Give your creativity full rein to soar. Seek out metaphors, images, mythical characters, role models, allegories, and/ or linguistic devices that express what really matters for you. Search for or allow this language to emerge from within as you explore your life and leadership vision.
- Be flexible with what kinds of vision you compose. Some people struggle with the idea of vision because they don't feel they can articulate a destination. Others simply prefer another type of vision, which is to view vision as a journey, or how you will travel through life. If vision-as-journey appeals to you more than vision-as-destination, that is fine.
- Try to include purpose, not just vision, in your poetic reflection. One way to differentiate between the two is to view vision as "what" (my destination) or "how" (how I will travel) and to view purpose as "why" (why I do this; why it matters).
- Try not to shortcut this process. Maybe it will take days or longer. If you use this exercise "only" to get started—that's okay.

Now that you have composed for a while, ask yourself by when you want to have made this future real.

STAGE 3: Recording
Record your vision in whatever form you choose. Perhaps you drew it and you are happy with the results. Perhaps you wrote a poem, a song, or a musical score and you want to work up a next version later.

Perhaps you want to lay down a musical track. Go beyond merely intending to record your vision to actually recording it.

STAGE 4: Performance

If you feel comfortable, share your vision with others. Doing so has the benefits of holding you accountable and providing feedback as you "demo" it out loud. Or, if you prefer, keep your vision to yourself for now.

In either case, schedule a review meeting with yourself in three months' time. The diary entry could read: "By today, I will have done X in service of achieving my three-year vision of Y." (Fill in X and Y. Adjust the three-year time frame as required.)

STAGE 5: Direction-setting

Direction-setting is a first move toward action. Direction-setting involves connecting multiple sources of meaning to your poetic reflection. The output from this stage is an outline AI transformation story. This outline will allow you—later and in your own way—to create a full transformation story that captivates others.

This stage involves translating what you visualized, composed, recorded, and possibly performed into an outline, "directional" transformation story by using the inputs below.

In their book, *Beyond Performance 2.0*, Scott Keller and Bill Schaninger argue that organizational change narratives optimally appeal to the following five sources of meaning:[13]

- Your company
- Society
- Your customer
- Your working team
- You personally

Great transformation stories draw on many of these sources of meaning. For each of the above sources, ask yourself what is relevant from the following categories:

- High and low points in your life

- Your strengths and weaknesses
- Your poetic visualization

For example, you may realize that being made redundant early in your career gave you empathy for how transformation impacts your suppliers. Keep reflecting on these sources until you feel satisfied with the narrative you have developed.

From the above sources, develop a long list of potential transformation story titles. Each title should potentially expand into a story that matters to you, one that is deeply personal in some way.

From there, choose one title that resonates most for you. Use your insights from this resource to develop your own transformation story outline that you share with colleagues.

NOTES ON CAPITALIZING ON THE FIVE STAGES

How you and your team use your stories will depend on many factors, including the type and stage of your organization's transformation and your digital context.

Following on from the team or departmental reflection exercise, consider who will regularly create holding spaces that will allow poetic and kinesthetic "larger will" interventions. Examples of such interventions include labyrinths, nature walks, and meditation retreats that incorporate physical activity. Such interventions constitute venues for embodied organizational awakening.

PAYOFF

Nudge others toward a more embodied culture.

Resource 19. Embodied Thinking Assessment

PURPOSE

- Encourage embodied organizational awakening.
- Champion multiple intelligences in the way you implement AI.

PROCESS

This resource takes the form of a short assessment that takes fifteen minutes per person to complete. After implementing the resource, collate, reflect on, and act on the results. Be sure to allow plenty of high-quality reflective time for your teams to digest and work with the results—weeks, months, or years if necessary.

Setup:

- To use this resource well, agree in advance the scope of the word *you* in the assessment. In this context, you might consider Dr. Otto Scharmer's four systemic levels: micro (individuals), meso (groups), macro (institutions), and mundo (ecosystems).
- Also agree in advance which parts of the chosen systemic level(s) you will assess.
- Create a simple online form or other questionnaire from the information below, ready to be distributed.

Assessment:

Numeric questions: Bring to mind the hardest problem that needed to be solved in the last year. On a scale of 1–7, to what extent did you use the following intelligences or types of thinking (see definitions on the following page)?

- Intuitive thinking
- Inductive thinking
- Deductive thinking
- Abductive thinking
- Mental pulsing
- Logical-mathematical intelligence
- Linguistic intelligence
- Spatial intelligence
- Naturalist intelligence
- Musical intelligence
- Bodily-kinesthetic intelligence
- Intrapersonal intelligence
- Interpersonal intelligence
- Existential intelligence

Free-form questions:
- How might you have solved that problem better if you had used other intelligences or modes of thinking?

- What are you assuming that is stopping you from using other intelligences or modes of thinking?
- What AI challenges may require you to use an intelligence or mode of thinking that you don't use much?

DEFINITIONS:

- Intuitive thinking: understanding things instinctively, without conscious reasoning.
- Inductive thinking: reasoning from the specific to the general.
- Deductive thinking: reasoning from the general to the specific.
- Abductive thinking: relying on inference to the best explanation.
- Mental pulsing: varying your mental focus from narrow task focus to broad big-picture thinking.
- Logical-mathematical intelligence: quantifying things, making hypotheses, and proving them.
- Linguistic intelligence: finding the right words to express what you mean.
- Spatial intelligence: visualizing the world in 3D.
- Naturalist intelligence: understanding living things and reading nature.
- Musical intelligence: discerning sounds, their pitch, tone, rhythm, and timbre.
- Bodily-kinesthetic intelligence: coordinating your mind with your body.
- Intrapersonal intelligence: understanding yourself, what you feel, and what you want.
- Interpersonal intelligence: noticing distinctions among others and sensing others' intentions and desires.
- Existential intelligence: the intelligence of big questions, such as "Why do we live?" and "What is love?"[14]

PAYOFF

- Solve intractable AI problems better and faster.
- Bring your organization's whole being into its collective thinking.

CHAPTER TAKEAWAYS

- Threshold leaders crystallize embodied intelligence in their teams and organizations.
- Threshold leaders transform their own and others' minds and attention by systematically accessing the wisdom that their bodies hold.
- This matters because teams and organizations with high embodied intelligence are culturally vitalized, ethically thoughtful, and impossible to control.
- Build a culture of multiple intelligences by valuing poetic and other existential input, cherishing embodied mental and emotional energy rhythms, and prizing knowledge that resides in bodies.

MATURING CONSCIOUSNESS

10

THE SENTIENT
VENDING MACHINE

*Just because there's a scary possibility out there,
it doesn't mean that we should assign a high probability
to that scary outcome.*

—CEO AND AI RESEARCHER BEN GOERTZEL

It would be impossible to write any meaningful book about leadership in an AI future without acknowledging and delving into the possibility of Artificial General Intelligence (AGI) and Superintelligence. What are the most important adaptations that leaders will need to make by the time AI usurps all our cognitive intelligence (i.e., in the Age of AGI), or also usurps all our other intelligences (i.e., in the Age of Superintelligence)?

In Path IV, I offer an agenda to help leaders shape a beneficial future even in these two ages. This agenda is necessarily speculative and even fantasy in some places, as we do not know how AI will develop, how far AI will develop and to what extent humanity and machines will integrate. My agenda is modest in that I offer it as a starting point on which others may build. I am confident of one thing, however: Humans have tremendous contributions to make.

Before coming to the proposed leadership adaptations themselves in chapters 11 and 12, this chapter lays a foundation by addressing the following questions:

- What is AGI and Superintelligence, and when might they emerge? Here, we explore definitions and speculation about potential future developments.

- What are the implications for leaders of an increasingly complex universe? Here, we encounter the core insight that, as AI drives increasing complexity in our world, threshold leaders are the ones who increase their own complexity of mind.

A SIMPLE TAXONOMY OF AGI AND SUPERINTELLIGENCE

Various definitions about AGI and Superintelligence abound in AI literature, and these definitions are not always reconcilable.[1] Where many AI writers conflate AGI and Superintelligence, I think it useful to separate them for one reason: AI will not likely achieve human levels of logical, mathematical, or other cognitive mastery (which relate in my taxonomy to AGI) at the same time as it also matches human levels of awareness and action in relation to emotions or existential issues (which relate in my taxonomy to Superintelligence).

Unlike some, I do not assume that intelligence equals cognition. Linguistic, interpersonal, intrapersonal, and kinesthetic intelligences include both cognitive and noncognitive elements. I therefore favor the following definitions:

- Artificial General Intelligence: "The ability to accomplish any cognitive task at least as well as humans." (Max Tegmark)
- Superintelligence: The ability to exhibit knowledge and skills cognitively, emotionally, spiritually, and physically at least as well as humans.[2]

By definition, AGI will be able to perform every single cognitive task that human leaders can perform. When you are in a board meeting, a performance review, a financial planning meeting, or a negotiation, how often do you use logic or mathematical skill? Quite often, probably. AGI will therefore be a seriously impressive and potentially useful collaborator. Also by definition, Superintelligence will not emerge before AGI emerges, as Superintelligence includes AGI.

AGI

When might AGI appear? I am assuming that AGI has not appeared anywhere so far. To the best of my knowledge, at the time of writing, this assumption is true. In the words of Canadian American psychologist Steven Pinker, AI systems are currently "savants," with "brittle mastery" of the problems they were set up to solve.[3] AGI is currently real only as imagined or conceptualized.

Beyond this, the following seems certain: At least one theoretical leap is required for AI to progress beyond today's narrow AI. In his chapter "The Limitations of Opaque Learning Machines," the computer scientist and philosopher Professor Judea Pearl argued that basic barriers stand in the way of what he calls *Strong AI*, which resembles what I define as AGI. As Pearl noted, "Current machine-learning systems . . . cannot reason about 'What if?' questions and, therefore, cannot serve as the basis for Strong AI."[4]

Pinning down a likely timescale for AGI's appearance is fraught with difficulty. Just because something requires a theoretical leap doesn't mean such a leap won't happen soon. We can describe human history in terms of a series of unexpected theoretical leaps. For example, Stuart Russell described how, on September 11, 1933, the problem of liberating nuclear energy went from impossible, in the opinion of many eminent nuclear physicists, to essentially solved in less than twenty-four hours.[5] And if money drives breakthroughs, then we may expect a breakthrough quickly. Companies such as Meta and DeepMind are investing heavily in pursuing AGI. In 2020, Google's parent company, Alphabet, spent $27.6 billion on research and development, an amount equivalent to the entire US Department of Defense budget and more than four times the size of the US National Science Foundation budget in the same year. Or perhaps a breakthrough will come from academia, where budgets are smaller and progress may therefore be slower.

We must be cautious about exact predictions. In 2019, I sat down with Professor Matthias Holweg, who leads the Artificial Intelligence Programme at Oxford University's Saïd Business School. We talked about AI as we sipped tea in the cool of a quad near his office. "With quantum computing plus pentabytes of data available on the internet, there will be no secrets," Holweg observed. "Add in some theoretical discovery or clever application of an old, failed method and, voilà, AGI." When Holweg said this, he immediately added, "But this is futurology. Where? How? When? We simply don't know."[6]

My own view is that we will not see AGI within the coming decade and that it may take many decades beyond that to come to the fore. This accords with Max Tegmark's observation that "virtually nobody" thinks that "superhuman AI" will appear "in a few years."[7]

Narrow AI breakthroughs will continue to be reported breathlessly, in ways that blur boundaries between science, psychology, and philosophy; boundaries I have sought to delineate in this book. I encourage you to read such reporting with a skeptical eye. Distinguish carefully between admirable narrow AI progress and as-yet-unseen theoretical leaps. Those who expect AGI to emerge within months or a few short years may be taking too seriously "the inflationary phase in the AI hype cycle in which we are living today."[8]

SUPERINTELLIGENCE

When might Superintelligence emerge? Here, predictions really are a fool's errand. The Age of Superintelligence will have started when AI matches or usurps all human intelligences, matching AGI while being emotionally sensitive and wise to boot. By this token, if Superintelligence were to appear at all, this could theoretically happen seconds, centuries, or longer after AGI emerged.

An elephant in this room is techno-humanism, where humanity and AI merge to the point where one cannot meaningfully distinguish whether it is the human or the AI that "has" a particular intelligence. In a world where humans and machines are highly integrated, defining identity is even more difficult than in a human-only context. Imagine a techno-human entity with all the love and wisdom of a human, combined with artificially supercharged intelligence. By my definition, this entity counts as super-intelligent. Given that, most likely, humans and machines will increase their levels of integration gradually in the coming decades, it will not be straightforward to point to a moment when Superintelligence emerged.

Techno-humanism aside, an AI that cannot *experience* all intelligences as humans do may still be counted as Superintelligent, as long as it demonstrates those intelligences to a sufficiently high level. In such a case, a Superintelligence would demonstrate or mimic emotional and existential intelligences functionally, without having acquired them in the sense of feeling what they are like.

To illustrate this point, consider BRETT (The Berkeley Robot for the Elimination of Tedious Tasks), who "has been folding piles of towels since 2011."[9] With the strapline or tagline "the robot that put some spunk into laundry," BRETT has caused significant excitement at Berkeley and beyond.[10] Of course, tidying-up guru Marie Kondo may beg to differ about the tediousness of folding towels. Quite plausibly, an AGI BRETT could very quickly design processes that outclass ours and commandeer any required capabilities from humans or machines. And a Superintelligent BRETT could design processes that also enable it to interact wisely and emotionally sensitively with others. In this Superintelligent scenario, BRETT would master body movement (and learned-about body movement), but would not feel what it's like to fold towels. (Insert "feel the boredom," "feel the joy," or "feel a sense of personal achievement," depending on your perspective on folding towels.) We are back to consciousness and qualia, discussed in chapter 1, and which I do not define as a necessary part of Superintelligence.

In other words, human leaders may retain some distinctive experiential qualities that even Superintelligence will not match. Threshold leaders cherish and emphasize these qualities. Tapping into a developing journey of mind and body is an essential element of leadership success as AI improves.

HOPE OR HOPELESSNESS?

I'm not arguing that there is no way that technological disaster will befall us. But I do contend that the wisest view is the hopeful one. Psychologist Andrew Bienkowski puts it this way, referencing the Greek myth of Pandora's box:

> Pandora opened a forbidden box that contained all the world's evils. When she lifted the lid to peek inside, the evils escaped. Pandora slammed the lid shut just as Hope, who was slower, was also trying to emerge, trapping it inside the box. The world had been a blissful paradise before Pandora's box was opened. Afterwards, suffering,

disease, and death plagued the land. Not until Pandora finally returned to the box and freed Hope did humanity find a way to survive suffering. Hope, we learn, is stronger than the evils of the world.[11]

I puzzled long and hard over claims made by respected computer scientists and others about AI dominating us in future scenarios. What I learned from researching AI literature was this: Just because it's possible that AI could spell danger for humanity doesn't mean that it is likely—for three reasons. First, the view that AI will dominate us rests on an assumption that advanced AIs will necessarily act from goals. However, goals are extraneous to intelligence. Young children illustrate this through growing and experimenting. Open-ended intelligence and "swarm intelligence" provide possible routes for AI to develop according to non-goal-directed paths. Second, even if AI does develop in a goal-directed way, these goals may well be collaborative, not competitive, with humans. Third, we must factor in human potential, not just AI potential. If you choose, you can be one of those who guides AI's path as AI gets more complex.

All this combines to mean that the leadership challenge of AI is adaptive, not technical, to use the language of Harvard University leadership lecturer Ronald Heifetz.[12] Technical challenges are those in which the skill set necessary to excel is well known, for example, the important and significant challenge of landing an airplane with a stuck nose wheel or applying a known natural language-processing solution to a new business sector. Adaptive challenges require more than just new technical skills; they also require a transformed mindset.[13] The adaptive challenge for leaders is to find ways to catalyze *both* advanced *and* beneficial AI.

Now, this AI challenge is really complex. AGIs won't just supply solutions, they will supply ever more difficult questions. And even if Superintelligences lack our soulful, embodied, generative humanity, they, too, will introduce more complexity to our society. Humans may also struggle to predict the future trajectory of self-improving, superhuman intelligence. So, the challenges of advancing AI will be astoundingly multifaceted.

It is true that extremism, discrimination, violence, and abuse in many forms indicate humanity's long-term failure to solve systemic issues. The causes of misfiring leadership are, of course, multiple. But solving adaptive problems is not beyond us. Every year, leaders solve highly adaptive problems, including problems of organizational and cultural transformation. As AGIs and Superintelligences introduce even more complex challenges, threshold leaders are the ones who will increase their complexity of mind to match and exceed this increasing technological complexity. Let's explore a thrilling developmental path that can support you to do this.

DEVELOPING CONSCIOUSNESS

In 2002, a colleague I'll call Josh stormed up to my desk, red faced, waving a printout of a PowerPoint chart in my face, shouting, "How could you do this?!" The previous day I got some analysis wrong, but—worse than that—I managed to get it wrong in a way that put Josh in a bad light. A bead of sweat rolled down my brow. I shifted uncomfortably in my seat, my unease heightened by knowing that I had been perfectly aware when I did the analysis that Josh would come off badly.

Josh tore into me. "Why?" he demanded. I was hurt, but in the words of the song "Cowboy Logic" by Michael Martin Murphy, I thought "if it hurts, hide it."

So, I hit back. "That's rigorous work," I said. And as soon as the words came out of my mouth, I could sense Josh's anger increase.

Of course, in hindsight, I should have said, "I'm sorry, I made a mistake. I feel awful about this. What can I do?" But I didn't. I just kept talking, and here he was scrunching my chart in his hand, veins throbbing, spittle flying.

Suffice to say that my follow-up of "Well, it just depends how you interpret the data" didn't help. This was my brittle self, sprinkled with a light dusting of a warped professional value of needing to be seen to be excellent. I refused to take a backward step, even though I knew I was wrong. In that moment, I was unreflective and had little capacity for merging my interests with Josh's or those of others. I was far from the threshold. By contrast with Ken, whom we met in chapter 2, I was profoundly decentered, neither rooted in the positive values of my employer, nor determined to inspire others in this situation.

It took me nearly twenty years to glimpse something of what I missed in that moment with Josh. Throughout the last half century, the American developmental psychologist Professor Robert Kegan has dedicated himself to researching, teaching about, and providing therapy based on a new field he pioneered: Adult Development Theory.[14] ADT is a multilayered field that deservedly enjoys influence among leadership practitioners.

At the heart of ADT is a stage model of adult development. Whereas forty years ago, the dominant model of mental complexity held that it increases in humans until the age of about twenty and then plateaus, ADT holds that adults can continue to increase their mental complexity in stages throughout their lives.[15] Each stage marks an increasing maturity in making sense of the world. Below, I describe four of Kegan's stages largely from a leadership point of view:[16]

- Teenagers and some in early adulthood feel embedded in their school and family as institutions of authority and role differentiation. They focus on keeping their body and its basic goals from disintegrating. This is akin to leadership as survival and the pursuit of comfort and pleasure (Kegan's "self-sovereign" stage).

- A leader's view of reality can then expand to embrace the values of a community, such as an employer, military unit, or religious community. The leader conforms to conventional norms and standards. This is leadership as expressed primarily through relationships, "schools of thought," or both (Kegan's "socialized" stage).

- Then, the leader again turns inward, finding new grounds for personal autonomy. Such leaders are reflective individualists, able to step back enough from the social environment to generate an internal personal authority that evaluates external expectations. This is leadership as the actualization of potential (Kegan's "self-authored" stage).

- Next, leaders can let go of being embedded in their own identity and embrace a culture of intimacy. This is a turning away from the self, back toward an integration with other people and with universal values. This is also leadership as thinking across longer time spans, holding divergent perspectives, and decreasing the clarity of connections between cause and effect (Kegan's "self-transforming" stage).[17]

The leaders who will flourish as AI improves (and especially as it approaches AGI or Superintelligence) are those who progress beyond the self-authoring stage toward the self-transforming stage.[18] Leaders who make this move step into an effective, yet liminal, threshold space. To bring Kegan's four stages to life in our context of threshold leadership, let's return to three stories introduced earlier in this book.

SELF-SOVEREIGN: MY REACTION TO JOSH

Back in 2002 with Josh, I was fragile and self-sovereign in many ways, a bit like a grumpy teenager in my twenties. I grasped poorly at a more socialized stage, trying to be rooted in the servant-hearted values of my employer. What I was doing was trying to simplify the world into a narrow definition of analytical rigor. Perhaps I did this as a coping mechanism against an increasingly complex business context, but, in any case, my leadership was more from the wolf pack than from the threshold.

The problem is that, as Kegan and Lahey astutely observed: "When we experience the world as too complex, we are not just experiencing the complexity of the world. We are experiencing a mismatch between the world's complexity and our own."[19] Many of us naturally seek to simplify the world, perhaps oversimplifying it as a coping mechanism. Threshold leaders are more aware than I was about what "has" them in its grasp. At the threshold, you embrace your own potential for growth by enhancing your complexity of mind so that it corresponds to the complexity of an increasingly advanced AI world.

I love the question *Would you follow you?* On sober reflection and with two decades' perspective, I know how I would answer that question of my wooden early 2000s leadership. Kegan and Lahey noted that "the story of mental complexity certainly does not end in our twenties."[20] I for one am grateful for that observation.

My favorite image of adult development is the image of a cross section of a tree trunk. Think of a tree's growth rings as representing not only your current stage of growth, but also the previous stages, which are still part of you. I like the idea that the rings aren't even. There may be smaller or larger gaps between the rings. I also like the idea that previous rings are valuable parts of the tree. Learning to love prior stages is part of the journey to the threshold. I learned a lot about blaming and the true nature of excellence through my encounter with Josh, and this learning stays with me to this day. This learning also inspires me to ask: Will we blame AI? Will we blame "our leaders" (whoever we view them to be)? Or will we increase our complexity of mind and create truly excellent solutions, however messy the collaboration?

FROM SOCIALIZED TO SELF-AUTHORING: MY COACHIFY EPIPHANY

By my late thirties, I had become socialized into norms of corporations, church, family, and friendship, to name a few. As I considered stepping away from Coachify, I was afraid of what I might find because my view of work realities had become largely oversimplified and largely limited to the world of start-ups. But I had started Coachify in part to produce an app that could be accessed by children living in urban slums, not just by business executives. At age twenty-one, I spent six months of my gap year working in the *favelas* of Jardim Olinda and Rebouças in São Paulo, Brazil. Two years into Coachify, I had become so decentered, fearful, and overwhelmed that I had forgotten this motivation.

In January 2017, I used the Honest Look resource and glimpsed a more self-authored form of leadership. As I turned inward and examined my assumptions, my awareness grew—awareness of what was going on for me in mind, body, and emotions. I connected thinking and being in self-authoring ways that hinted at threshold leadership beyond.[21] I started to articulate my own expectations outside just what my investors or customers wanted. As a result, I crafted a personal mission statement that better reflected my whole self, not the self I thought I was supposed to project to the world. I emerged with a more embodied, self-authored, near threshold sense of feeling alive, from which I generated a vision that proved productive for the next few years.

DEEPENING IN SELF-AUTHORED AND GLIMPSING SELF-TRANSFORMING: BEATRICE

In chapter 5, we met Beatrice, a US-based senior lawyer who broke through to higher performance and satisfaction, in part via thinking independently. Beatrice also made an important developmental step during our coaching. Previously, she regularly protected herself by remaining emotionally distant from colleagues, who saw her as aloof. Beatrice told me that she usually created this distance in one of two ways. "When I'm working a case and something difficult happens, either I withdraw and take my hands off the steering wheel completely, or I push push push to get around or through whatever is in the way." Leaning back and looking to the sky, she added, "I'm either absent or monstrous. People avoid me."

I invited her to play with moving from "I am this" to "I have this" in relation to various issues she raised. For example, she moved from thinking, "I am aloof" to thinking, "A part of me behaves in aloof ways sometimes." This was a classical developmental move rooted in what Kegan calls a subject-object shift. Earlier, Beatrice was subject to withdrawing. Later, she realized that a tendency to withdraw is something she has, not something she is. She was even able to talk to her aloofness and try different things with it. Withdrawing became an object to her, not a core part of her being, and she became more conscious and powerful as a result.

Beatrice made the move from subject to object, seeing distance as something she "has," not something that "has" her. This move was powerful for Beatrice because she didn't need to jettison distance entirely from her leadership armory. She now allowed herself to create a little distance in some settings, such as in a tough performance review conversation. This was a firm move toward threshold leadership, as she decreased the role of distance in her life while still using it occasionally.

Beatrice displayed other threshold tendencies. She improved her ability to hold tensions between the competing needs of different people in her life and started to evolve a richer life purpose that connected more with issues of planetary importance. Beatrice demonstrated what Kegan described a quarter of a century earlier: "Adulthood itself is not an end state but a vast evolutionary expanse encompassing a variety of capacities of mind."[22] Accessing this expanse will be key in the Ages of AGI and Superintelligence.

As I look back over my twenties, thirties, and forties, I see a trail of discomforts that arose when something did not match with my current beliefs. Previously, I mitigated this by assigning to someone else responsibility or blame for "terrible" actions: James Watson wasn't interesting enough, Josh didn't clarify his request, Oliver walked away from Coachify too abruptly—you name it. Today, I view this discomfort with interest, curiosity, and learning. Threshold leaders act similarly. In the Ages of AGI and Superintelligence, leaders can expect regular chaos and paradox. Threshold leaders will be at ease with this, even if they feel quite lost from time to time.

It turns out that the three pathways already presented in this book are also developmentally mature pathways. The more you cultivate stillness, the more you support leaders to face paradoxes between their true and false selves. The more you nurture independent thinking, the more at ease you are with partial or incomplete systems or organizations, including your own systems of knowledge or progress. The more you embody intelligence, the more you see that progress happens through a dialectic between multiple selves, including feelings, body, and mind. As a result, mature threshold leaders cultivate space, not exhaustion; generative, not directive environments; and multiple, not singular intelligences.

THRESHOLD RESOURCE

RESOURCE FOR ANY LEADER

Resource 20. Honest Look
PURPOSE
Identify and let go of what blocks you from increasing your maturity.

PROCESS
One of the most common leadership mistakes is to think that what got you here will get you there. (Wherever "there" is, and it's rarely the same as "here.") In the Age of Superintelligence, effective leaders will cultivate an inner life that helps them deal with the undulations of leading in an increasingly complex future.

This developmental resource takes the form of a retreat exercise.

This exercise will help you, in the words of Parker Palmer, to "get off your white horse of ego and take an honest look at your liabilities and at the truth of who you are." I developed this exercise based on some pages from Parker J. Palmer's exceptional book *Let Your Life Speak*. Some prefer to break this one-day retreat exercise into four smaller elements, each approximately ninety minutes long.

However you use this resource, I encourage you not to rush any step. Deep wells of growth reside in every one of them. Try to avoid the temptation to read ahead to future steps. Be as present as you can to each part of the reflection. Times are approximate.

PLANNING THE DAY
In advance of the day, book a call with someone who knows you well, preferably someone who listens well and encourages you. The purpose of this call is for you to explore thoughts that emerge for you during the day with someone you trust. Schedule the call for the afternoon of your retreat day.

Seek out a quiet, restful place for your retreat day. This could be the beach, a retreat center, or a national park—any place you feel at ease and that inspires you in some way. If you can afford it, stay overnight. Consider this an investment in your best self.

Plan to detox from digital devices between 9:30 a.m. to 4:00 p.m., apart from the time when you will make the call mentioned above. By detoxing in this way, you stand the best chance of creating the liminal, high-quality space you need to benefit most from this exercise.

To aid your reflection, you might like to use the poem, "A Poem For Someone Who Is Juggling Her life" by Rose Cook, available in her book *Notes from a Bright Field* (published by Cultured Llama).

On the day itself, bring this book with you, together with a notepad and something with which to write and/or draw. Intersperse the recommended steps with walks outdoors. Alternatively, perform all the steps outdoors.

STEP 1 (MORNING): SETTLING IN
Start by enjoying a cool drink, a mug of coffee, or a cup of herbal tea. Then check in with yourself as follows:

- Write in your journal whatever is on your mind. Not necessarily a to-do list, but all the things that are front of mind for you. The purpose of doing this is to connect with the reality of where you are, at this moment.
- How are you feeling? Write out as many emotion words that come to mind.
- Set an intention for your retreat day. Pray and/or meditate as you wish.

STEP 2 (MIDMORNING): MY STRENGTHS AND I

Ask yourself, "Is what I've done my life?" Reflect on this question for at least twenty minutes. Record your reflections through writing, drawing, or other means.

Reflect on what is lovable about you. What are your strengths? Write this out. Resist the temptation, at this point, to move on to what is not lovable and not a strength.

If you haven't already, now list the ways in which you view yourself as intelligent.

STEP 3 (LATE MORNING AND OVER LUNCH): THE WINTER OF MY FALSE SELF

Now it is time to lean into the ways in which you have lived winter. In many ways, this is the core of this whole retreat day and may feel like hard work.

List moments of darkness from your life. These moments can include any kind of upset, however small or large. You might call to mind an occasion at work, home, or elsewhere, where you felt aggrieved by something someone did or omitted doing. Some of the "moments" may be periods of time, spanning days or weeks. If you are thinking of a moment that is longer than that, try breaking it down into smaller chunks, for the purpose of this list. At this point, do not judge, analyze, or otherwise scrutinize what is on your list. Generate as many "moments" as you can.

Go through your list. Ask yourself, "What other moments of darkness can I recall?" Keep asking yourself this question until you are fairly sure that you cannot generate any more moments.

After completing the above, cast your mind back to each moment. For each one, note ways in which your best self showed up in that moment, and ways in which your worst self showed up. This might include ways in which you contributed to what happened, for example, showing (best self) humility or (worst self) lack of care for another person. Put differently, in what ways did who you really are emerge in each moment, and in what ways did that moment in your life trigger your shadow side?

Ask yourself, "In what ways was I embedded in my own identity earlier in my life?"

Over some of your lunchtime, continue to reflect on the above. You have started a journey of turning away from the self, back toward an integration with other people and with universal values. Also allow yourself time during lunch just "to be"; in other words, not reflecting deeply on the same topics all day. Let your thoughts wander.

STEP 4 (EARLY AFTERNOON): LETTING GO

Read this story of Bill Plotkin's loyal soldier, as told by Brian Draper.

Plotkin wrote about Japanese soldiers in World War II who survived on their own in remote regions after being shipwrecked or shot down. On being discovered after the war was over, Plotkin explains, "They were told the war was over, but this was literally unthinkable to them: the war could not be over because their loyalty to the cause was what had kept them alive all those years."

These "loyal soldiers" ultimately required compassionate help to stand down, and as part of the process were welcomed home with honor, gratitude and loving-kindness, sometimes through ceremonies in which they were encircled by their communities and thanked over and over for what they'd done. Plotkin suggests that we each have a loyal soldier—what we might call the voice of our ego—who has, in effect, been hiding out within since an early age. It's been fighting unswervingly on our behalf, through our childhood and beyond, to keep us, our image and our place in the world safe.

The problem is that the outlook of the loyal soldier comes at a cost. In the case of the Japanese, they welcomed the soldiers home, thanked them repeatedly for their service, repeated that the war was over and helped them, in time, to find new roles in society.[23]

Reflect on the notes you made beneath each item in your list of moments of darkness. Start to amalgamate themes, to identify the loyal soldier that sometimes trudges, sometimes marches, to the forefront of your life. Acknowledge that your false self has been necessary for a time but is less helpful now. Write down the ways in which it was helpful and therefore still remains a part of you. Also articulate what your false self has cost you.

Next, dare to let your false self fall, at least a little. This may take courage. One way to do this is to identify one thing you would like to let go of, to leave symbolically behind on this day. Find something heavy near you, like a stone. This object should not be something you brought with you to this retreat day. Also find something light, like a piece of paper or a blade of grass. Feel the weight of the heavy object and imagine this is what you've been carrying. Now, pick up the light object. Which would you rather carry on the road ahead? Now take the heavy object. Hold it and imagine it's the thing you'd like to leave behind. When you are ready, set it down and decide not to pick it up again.

STEP 5 (MIDAFTERNOON): SYNTHESIS—TOWARD THE SUMMER OF
    YOUR TRUE SELF

Review your listed reflections of moments of darkness. Next, attempt to characterize something of your true self—who you really are and who you really are becoming. This will be a first draft. As you start to think about this, try to hold different perspectives, as anyone's true self also includes a multitude of selves.

Your first draft will be an early draft of your "true self statement." The following questions may help your reflections:

- What makes you feel fully alive?
- When do you feel fully embodied?

- What brings you joy?
- Where does your deep gladness meet the world's deep need?

Take your time. Start with a blank sheet of paper. Your draft could be in the form of a drawing, poem, story, bullet-point list, opera, diagram, or whatever you find inspiring. Mine was in the form of a life mission statement with poetic elements. Get creative, logical, illogical, intuitive, or detailed, as you prefer. Use the power of your imagination.

Share this draft with someone on a call. Listen to their comments. After you finish the call, check in with how you feel about your draft "true self" statement. Modify your draft in whatever way you feel appropriate.

STEP 6 (LATE AFTERNOON): FINISHING WELL

Make this true self statement real in your life, first by visualizing it in your mind. Remember, you are on a leadership edge, and it's okay if not everything lines up neatly.

Begin by closing your eyes and picturing your life in two years' time when you are fully living as your true self. Take five minutes to imagine this self, thirst for it, and delight in the fact that you have it in your mind.

Now, spend a few minutes visualizing your best next steps—steps to move in the direction of the future you just pictured. Andrew Bienkowski wrote, "It requires great courage to choose to be your most authentic self when others all around you are acting and pretending to be what they think is expected of them." How will you contribute in the world of tomorrow? If you think it appropriate and if you feel ready, begin to make commitments that flow from your true self statement.

Record any actions or other reflections that come to mind.

Close by writing three things from the day that you are grateful for.

AFTER THE DAY

Share your draft "true self statement" with your family, colleagues, coach, mentor, and/or a friendly dog.

PAYOFF

- Thrive in the future, no matter how techno-humanistic that future becomes.

- Become a more effective AI leader by learning deeply from upsets.

CHAPTER TAKEAWAYS

- At some point in the coming years, AI may achieve general cognitive intelligence (AGI) and may subsequently master all domains of intelligence (Superintelligence).
- Leaders who embrace complexity and development move into an effective, liminal, threshold space.
- At the threshold, progress toward a self-transforming form of mind. This move will help you match, work with, and possibly exceed the increasing technological complexity that the two future Ages of AGI and Superintelligence may bring.

11

KNEELING AND FREEWHEELING

People, ideas, machines—in that order.

—COLONEL BOYD

In 2001, I watched a video clip of a British business leader in the manufacturing sector, whose competitor had just discovered a major product defect. The leader ordered a copy of his competitor's product, tasked his factory leaders with finding and fixing the defect, and sent the solution to his competitor. Whatever you may think of the impact on shareholder value from this leader's action, I view it as a highly conscious, loving, threshold response. This leader expanded the space of what was possible in his industry. As AI improves, such maturity will be even more vital if we are to thrive.

In the Ages of AGI and/or Superintelligence, threshold leaders can usefully adopt the following five mature qualities:

1. Humility: Prizing encouragement and love of knowledge over love of being right or best.
2. Ease with tension: Being comfortable with paradox; actively including humans, machines, and techno-humans who ask uncomfortable questions.

3. Play: Enjoying learning and exploration voluntarily for their own sakes.
4. Love: Moving beyond fear by bringing your whole self fully to others.
5. Wisdom: Respectfully approaching mystery, embracing an emergent universe.

The first three qualities are discussed in this chapter, being human qualities that AGI machines won't match, but that Superintelligent machines will match. The final two qualities are discussed in chapter 12, being human qualities that neither AGI nor Superintelligent machines will match.

The five qualities do not include leadership disciplines such as planning and accounting, since AGI will excel at them. I selected the five qualities using three criteria: breadth of intelligence, effectiveness, and developmental edge. First, all five qualities surpass merely cognitive (logical-mathematical) intelligences. Second, relevant studies associate humility, ease with tension, and play with higher leadership effectiveness, and these three qualities map well to what is needed to create beneficial AI in the Age of AGI.[1] Third, the final two qualities sit on the highest developmental plateaus.

The five qualities are relevant for you in today's Age of Narrow AI, not just tomorrow's Ages of AGI and Superintelligence, whatever your organization's size, sector, geography, and stage of transformation. Especially in future Ages in which AI can outthink humans and even relate to humans better than we can, leaders will have little else to distinguish themselves apart from such "threshold" qualities. At this threshold you can dance at the edge of what it is to be human, even what it is to exist. The five qualities are our best hope, our portal to flourishing in our paradoxical futures.

HUMILITY

Humility may be defined as having "an unusually low concern for status coordinated with an intense concern for some apparent good."[2] This often countercultural quality will become vital as AI improves. Consider the following example provided by philosophers Robert Roberts and Jay Wood, in which they quoted from Mariano Artigas's and William Shea's book, *Galileo in Rome*:

> As brilliant and productive a scientist as Galileo Galilei was, his work was impeded by his arrogance. . . . He overestimated the probative force of his arguments for heliocentrism, and thus underestimated the justification of those who hesitated to accept the hypothesis. In fact, his favorite argument—that the earth's motion accounts for the tides—was unsound. "To the end of his life, Galileo held to a simplified version of the Copernican system in which all the planets move in perfect circles. Although he preached open-mindedness, he never lent an ear to Kepler's arguments about elliptical paths."[3]

According to Copernican heliocentrism, the earth rotates daily and revolves around the sun. But alongside this revolutionary and accurate idea, Galileo also clung overconfidently to some unsound views. His contemporary Orazio Grassi commented that "Galileo caused his own ruin by thinking too highly of himself and despising others."[4]

Even a leading scientist like Galileo could learn from those around him. In his arrogance, Galileo underestimated factors that could have elevated what he contributed to the world. Had he kneeled in humility, metaphorically speaking, who knows what more he might have accomplished.

As AI improves, our humility can impact our whole universe, not just our backyard of Earth and Sun. AGI may surpass our ability to set performance aspirations and select an appropriate organizational transformation approach. Threshold leaders complement this by balancing confidence with humility, thinking not just with the head but also with the heart. In an Age of AGI, threshold leaders will show two kinds of humility:

- Epistemic humility: Prioritizing love of knowledge more than love of being right or best
- Competitive humility: Encouraging others and avoiding cabals

We will consider each type of humility in turn.

Epistemic Humility

Epistemic means "relating to knowledge." Epistemic humility, therefore, includes the two notions that our knowledge is always provisional and that we love ideas that are not our own.[5]

Threshold leaders are keenly aware that their knowledge is transitional. AGIs will inevitably develop knowledge far beyond ours. Vastly increased amounts of data and processing power may mean that "we" (being humans and machines summatively) "know" more, but human leaders may rarely know how machines came to a given answer, so humans may "know" little that is relevant. In such a context, threshold leaders will adopt a curious, open stance toward knowledge, welcoming inputs from AGI that are potentially beneficial for our species and universe. With Josh in 2002, I fixedly believed my ideas and knowledge to be complete and final, the worse for my satisfaction and leadership performance. The less we pretend that our knowledge is relatively complete, the less we will cause our own ruin.

I find it exhilarating that machines will know more and more. Imagine if AGI rapidly experiments and infers deeper truths about our origins or what makes us flourish. How exciting! And how risky, if we guide AGI in brittle, fixed ways.[6]

This is where the second—uniquely human—kind of epistemic humility comes in: Threshold leaders love ideas that are not their own. At our best, we find joy and lack of ego in new ideas. One thing we can be sure of in the Age of AGI is that new ideas will dominate the landscape. It stands to reason that the more we imbue emerging AGI with human values and culture, the more likely it is to operate in harmony with those values and cultures, when it surpasses us in more abilities.

As a result, let's imbue a love of others' ideas in everything we do and—as far as is wise—in what AGI does. This is a brave place to put ourselves. Where some media sources build leaders up, threshold leaders embrace (also) being ground down, allowing their most deeply held ideas and beliefs to be challenged. At the threshold, you find the courage to do this.

Competitive Humility

As AGI advances, it is not inconceivable that a very small group might be given significant power to determine humanity's direction in relation to AI.

Already much data and AI expertise are highly concentrated in relatively few corporations.[7] Ben Goertzel summarized the views of influential voices who favor this future as follows: "A few brilliant, right-thinking mathematicians and philosophers locked in a basement are most probably our best hope to save humanity from the unwitting creation of Unfriendly AI by teams of ambitious but not-quite-smart-enough AI developers."[8] Goertzel disapproves of this future, rightly so in my view. What a revolting future! If the "few" are in control, the diverse contribution of the rest of humanity gets diluted. If democracy is the least-worst form of government, what worse future may we create than a plutocracy of a few mathematicians, data scientists, and AGIs?

Many ages may be upon us, including the Age of AGI, the Age of Superintelligence, the Age of Surveillance Capitalism, the Age of Pandemics, the Age of Polarization, and the Age of AI Warfare. As global issues get more complex, threshold leaders increase their complexity of mind in part by prizing humility in a competitive context.

Why is humility relevant in the context of competition? A reason often given for AI's supposedly likely dominance of humans is that machines will outcompete us, so surely, some might argue, we should respond in kind. However, competitive goals need not govern our human-AI progress. Beware assuming that competition (where each "side" tries to outthink the other) will prevail over collaboration (where our thoughts, decisions, and actions are more congruent with our shared sense of self).

Success does not always flow from competitive goals. Take two examples: First, effective early responses to the COVID-19 crisis depended on collaboration in vaccine development, contact tracing, marketing, and manufacturing.[9] Second, Professors Nicolaj Siggelkow and Christian Terwiesch advocate four business strategies to respond to seismic technological shifts. Two of these strategies ("respond to desire" and "curated offering") represent a collaboration between company and customer.[10] Especially as AI improves, the most effective leaders will prize collaboration.

Some may argue that, in "normal" times, leaders collaborate only insofar as it helps them accomplish something more fundamental: competition, even self-serving competition. However, why always compete? The big mistake we often make in over-indexing on competition is where we assume that outperforming my neighbor means that I must be any good. In fact, at an individual level, outperforming my neighbor does not necessarily mean that I outperformed a single one of the other 7.999999998 billion people on the planet.[11]

Today, being a senior leader often carries with it status elevation and associated status symbols, such as a managed social media account, a gold-plated health plan, or a high-tier company car. Threshold leaders know that such elevation and symbols are transient. Instead, they humbly emphasize the unimportance of status symbols.

In addition, threshold leaders compete humbly by prioritizing open, transparent, collaborative intelligence. They know that credible models exist for human–AI collaboration that will serve us well into the future. For example, Hannah Fry and Garry Kasparov describe examples of integrated human–machine collaboration in healthcare, sport, policing, the judicial system, and online shopping.[12] In these areas and others, self-transforming threshold leaders "recognize their commonalities and interdependence with others," including with machines.[13]

Competitive humility also includes a shift toward AI coordination, where we install disciplines of planning and control to help us understand, predict, and mitigate the impact of advancing AI technologies. Stuart Russell offered the Food and Drug Administration as a partial model for such a coordinating body.[14] Although the FDA is not perfect, it has played a valuable role in protecting public health. To the extent that we regulate AI in areas, such as in synthetic biology or biohacking, I advocate encouraging a broad range of voices rather than tending toward a narrower concentration of expertise. Through such coordination, or competitive humility, we might for example broaden international consensus that a chat-bot should announce that it is not a real human.

Such openness and coordination matter if we envisage AGI playing a meaningful role in politics and policy, as it surely increasingly will. In 2019, an AI-powered virtual politician named SAM was slated to join the general electoral race in New Zealand. What would you think if SAM stood for office in your neighborhood? Now that AI can be used to fake videos and other forms of evidence, where do you think we should draw the line between justice and privacy in regulation? How do we minimize abuses of power in government and companies, if surveillance increases? We do not have to generate AIs that control, subjugate, and eliminate us. But to involve AGI well in politics and policy, it will serve us well to promote consistently open debate among diverse voices.[15]

Progressing through Kegan's developmental stages typically involves increasing epistemic and competitive humility. As you move from the self–authoring stage to the self-transforming stage, you start to recognize that you are not "had" by your values, priorities, or objectives—rather, you "have" them so you can hold them up and contemplate them. When you are in the self-transforming stage, you can say that these values, priorities, or objectives may all be just temporary. You can say that you may achieve more information and better insight about inequality, competition, and privacy later on.

As a result, self-transforming threshold leaders know that their ideas are provisional. They realize that they will never be done honing their internal value system, their internal model of the world, and therefore they love the insights and practical ideas that others contribute, even where these contributions jar. They have a humble awareness that their leadership will always be a work-in-progress.

EASE WITH TENSION

In the Age of AGI, threshold leaders will be at ease with tension in their own thinking and in that of those around them. It's long been advocated that holding tension in thinking is a good thing. In 1890, Thomas Chamberlin proposed the idea of "multiple working hypotheses." In 2011, Professor Daniel Kahneman wrote on cognitive ease, recommending that we think through again what, previously, we were inclined to believe simply because it has been repeated so often.[16] And in 2016, Robert Kegan and Lisa Lahey noted that mature, developmental organizations work with tension in a nuanced way, considering destabilization to be constructive and viewing error and weakness as opportunities and assets.[17]

Some might argue that AGI will be at ease with thinking-tension, on the basis that even today's AI systems (let alone those of tomorrow) have internal tension built in. The technical terms for such systems are Competitive Adversarial Systems (CATs) and Generative Adversarial Networks

(GANs).[18] For example, AlphaGo learned via constant friction between two types of networks that challenged each other.[19] Surely, the argument might go, if today's AI systems use tension this well, then tomorrow's cognitively masterful AGIs will do even better. While this may be true, three reasons stand against AGIs becoming as easeful with tension as humans.

First, the friction (or tension) within today's CATs and GANs operates only at a functional level, not an intentional level as with humans.[20] "Functional" refers to the fact that AI networks optimize against an objective or objectives that humans provide. In the case of AlphaGo, scientists set the objective of winning at Go, and AlphaGo duly obliged. Competitive AI systems do not generate intentional purpose. Only humans provide this, which is why AI systems can't yet manage conflicting human preferences for which they were not trained.[21]

Second, "ease" differs from "challenge"—the former being a quality and the latter a capability, with only the latter, it seems, foreseeably open to machines. Threshold leaders do not know how tense and emergent stories will end, and they are at ease with this fact. This ease helps them take on a wide variety of perspectives in a nonjudgmental way and helps them unearth value in different viewpoints. Let's illustrate this point with the example of climate change. AI already assists us in addressing this challenge, by predicting extreme weather events, modeling emissions, optimizing traffic flows, improving building energy consumption, and producing climate models. But tricky human-related tensions inhabit the spaces within and between these and other factors. For example: What ecological price are we happy to pay for the certainty that familiar human-led solutions provide? Any route forward in relation to climate change will likely lead to loss for some and gain for others. This reality frequently prompts emotionally heated discussions. Alone, AGI will be ill-equipped to navigate this terrain, as even cognitively masterful machines will neglect key emotional and other complex resonances that these knotty debates entail.

The third reason AGIs won't manage tension in thinking as well as humans is that, at our best, we humans find gifts in upsets in ways that help us navigate our most difficult moments of tension. The heartrending yet uplifting story of Andrew Bienkowski illustrates this deeply human quality.

Bienkowski was six when he and his family were exiled by the Soviet Union from Poland to Siberia. Seeing the lack of provisions available to the family, Bienkowski's grandfather made the excruciatingly difficult decision to consume less of the meager rations available to him, thereby increasing the chances that the rest of his family would survive. Eventually, Bienkowski's grandfather died and the family gained permission to bury him on the outskirts of the town. The ground was frozen solid, so they weren't able to bury the body as deep as they would have liked. They dug down a little, as far as they could, and spread some rocks around the gravesite.[22]

Returning the following spring to pay their respects and bury him properly, the family realized with horror that the grandfather's body was spread out with bones everywhere. Wolves had got to him. "Ever since that sad, painful day on the plain," Bienkowski wrote, "my grandfather has been inextricably linked in my mind with wolves." At such a difficult time, with his mental and emotional energy so low, there must have been great temptation to rage and to hate wolves for the rest of his life, because of the disrespect they showed to his grandfather.

Instead, Bienkowski wrote one of the most inspiring paragraphs I've ever read. "With greater understanding," he began, "I have developed an extraordinary fondness for wolves as well as compassion for their plight—creatures living by their wits on the fringes of land that unfortunately often overlap with human ranch-lands. The life of a wolf is not easy."[23]

People are not wolves, although we sometimes tear into one another from time to time. I don't know where Bienkowski found it in him to view wolves in this light, but he powerfully showcased maturity. Even in the most difficult circumstances, Bienkowski found sympathy and concern for wolves, drawing on reserves of courage and ease that will elude AGI. Somehow, Bienkowski resolved tension in his heart, finding gifts in an unlikely place.

I am reminded how amazing it is when we encounter someone who stands on the holy ground of grief and pain, yet speaks with their whole being about things like joy, life, and laughter. During the COVID-19 pandemic, I heard many stories of people finding greater presence in blessings, knowing the reality that they could be taken away at any moment.

In the responses of Bienkowski and those affected by COVID-19, what is going on is an ease with rupturing false binaries. For example, Bienkowski ruptured the false binary of "wolves good or bad." Threshold leaders cultivate ease with cracking open flawed dichotomies around them.

We will need this ease much more as AI increases in scale, scope, complexity, and skill. At the threshold, leaders cultivate ease with the most difficult tensions, knowing in their hearts the words of the artist Sophie Hacker: "Suffering is not something that happens *to* us, but *in* us."

At the heart of this is a deep threshold quality of taking multiple perspectives. In their excellent article "Understanding the Leader's 'Identity Mindtrap,'" Jennifer Garvey Berger and Zafer Achi recommended that leaders regularly ask themselves, "How could I be wrong?"[24] This question has value not because it makes "your beliefs bulletproof but rather . . . open[s] them up so that you recognize other ways of seeing the world that might be helpful to you . . . [and] opens us up to new possibilities."[25] Taking multiple perspectives is one of the best ways to cultivate ease with tension.

The first two sections of this chapter (humility and ease with tension) interconnect. The British mid-twentieth-century academic C. S. Lewis coined the term *Inner Ring*, which Alan Jacobs explains as a group that "discourages, mocks, and ruthlessly excludes those who ask uncomfortable questions."[26] While Inner Rings address tension superficially and easily, healthy communities face difficult questions head-on. Inner Rings are found in organizations the world over, including in government, technology companies, investment firms, and science labs. If we are not careful, AGI may seduce us into creating more Inner Rings. Against this, threshold leaders humbly invite others in, rooting out and resisting Inner Rings.

PLAY

Picture the scene: I'm sitting in my home office on my wheeled, posture-kneeling chair, fingers poised above the laptop, about to finish writing an early draft of this chapter. My (then) three-year-old daughter, Phoebe, bursts into the room. Any initial hint of frustration at being interrupted is washed away by the sight of the huge smile on her face and her big eyes sparkling at me. Within thirty seconds, Phoebe has boarded my chair, and we're free-wheeling around my office, arms out, whooping in delight. Anyone who claims that people who work from home don't get much done may have a point. Or do they? After a few minutes, Phoebe toddles off to the other end of the house to find Mummy, and I'm back at my desk, working. The wild joy that I feel proves productive. After Phoebe leaves, new thoughts emerge;

I feel fresh, and my reflection is deep. My writing during the following forty-five minutes was some of the best and most creative I had done in weeks.

This section is about exploring how play helps leaders as AI improves. I define play as enjoying learning and exploration voluntarily for their own sakes.[27] While drafting this section, I was amused to see that my iPad autocorrected "ludic" (the adjective from play) to "ludicrous." For some, the idea of play being relevant to leadership is indeed ludicrous. Surely we should rather do something useful?

Yet psychologists note that play is developmentally useful, including for leaders.[28] We also know intuitively what play is, as we remember it from our childhood. Even Anne Frank, Andrew Bienkowski, and others brought up in horrifically difficult situations view play as a vital part of their early growth.[29] Play can take many forms: music, dance, art, thinking, and even humor, though not the destructive kind. And it goes deeper than that, to simply being caught up in the present, riffing, experimenting, being wild, being in flow, and allowing joy to course through.

Those playful moments with my daughter helped greatly. I'm glad I allowed Phoebe in. The heart of our encounter could not be rendered in code. In one sense, this freewheeling differs greatly from the kneeling posture of humility we discussed earlier in this chapter. But in another sense they are similar, as both tap into our humanity in a way that AGI won't match. In the rest of this section, we explore how play increases creativity and inclusion.

One of the major mistakes that many people make when it comes to AI is to assume that providing options (which AI already does) is the same thing as being creative (which AI does not yet do as well as humans). This is demonstrated perfectly by Autodesk's Dreamcatcher AI, a tool designed to "enhance the imagination of even exceptional designers" and to "heighten creativity": Throughout a design process, this AI performs the myriad calculations needed to ensure that each proposed design meets the specified criteria.[30] This frees the designer to concentrate on deploying their uniquely human strengths such as professional judgment, aesthetic sensibilities, and playing with options that Dreamcatcher provides. Dreamcatcher mimics elements of human creativity but fails to bridge the gap to humanlike sparks of inspiration.

Such a gap will endure into the Age of AGI when some human faculties will remain unrivaled. As Dr. Stuart Brown, psychiatrist and founder of the National Institute for Play, explains, "Play shapes our brain, helps us foster empathy, helps us navigate complex social groups and is at the core of creativity and innovation."[31] In addition, some leading Chinese and Japanese artists, as well as many first-rate scientists, sit back and playfully connect what wants to emerge.[32] Play is vital to their excellence. Threshold leaders will use play to explode creative color into otherwise grey strategizing, even in an age where machines are exceptional at generating and sifting options.

The sober reality is that play will help you onto the threshold. Garvey Berger and Achi put it well: "The self-transforming mind offers an automatic lightness that is one of the hallmarks of its state—life is serious and not serious at the same time."[33] A desire to play or laugh isn't something that algorithms feel, if they can be said to feel at all. An AGI will not remember what it was like to play as a child. In these differences lie huge creative possibilities for you as a leader.

Play also matters in an Age of AGI because play is universal. In a world where inequalities are increasing and are set to increase, play is the great leveler. Since the COVID-19 pandemic forced many meetings online, I've lost count of the number of occasions where a cat or child entered a virtual room containing a group of leaders from different levels of seniority. As someone on the call started to play, everything changed. The ease, the empathy, and/or the laughter equalized roles and brought a groundedness and then a joyful focus to our discussions. Unless we are careful, advancing AI will exacerbate economic, social, and other inequalities. If you believe that equality is important, step onto the threshold and champion play.

Put another way, play is universally generative. One 2017 study of three thousand adults found that playful people are good at observing, can easily see things from new perspectives, and can turn monotonous tasks into something interesting.[34] These benefits are fueled by the physical attunement, connection, and joy that arises when people play together.[35] Simply put, leaders create more by leading playfully.

What does it feel like to be with a playful leader? It feels like savoring life, like filling your day with many different enjoyable roles. This is no soporific or mindless 24/7 screen time, as depicted in the film *Wall-E*.

But there's something fundamentally ossifying about a concept of life and leadership that excludes play.

Threshold playful leadership is spirited, elusive, and unpredictable, like a kite at Key West. Threshold leaders grasp at something unseen, something deeper. Ultimately, becoming a playful leader is a choice you have. If you make this choice, you may become less wrapped up in your prior views of success. And you will likely flourish and help others flourish where it matters most.

THRESHOLD RESOURCE

RESOURCE FOR LEADERS IN LARGE ORGANIZATIONS

Resource 21. Influence Growth in Maturity
PURPOSE
As AI moves toward AGI, shift your team or organization's culture toward leadership maturity, especially via humility, ease, and play.

PROCESS
Use these prompts during a multi-month or multi-year transformation process.

Influencing growth in any era is a complex task, as cultural transformation is involved. In the Age of AI, this task is even more complex. This resource is not a holistic strategy or complete program for cultural transformation in an AGI context. Rather, this resource contains prompts to help you turn your workplace into an arena where growth and development can enduringly happen in a techno-human future.

This resource takes the form of questions and comments structured around the four-part structure of the Influence Model.[36] Applied in the AI-relevant context of this chapter, the four parts are:

1. Building understanding of and commitment to a humble, easeful, and playful culture.
2. Establishing processes and systems to support this culture.
3. Role-modeling this culture.

4. Establishing skills and training to support this culture.

A good way to use this resource is during leadership team meetings or departmental summits. Ensure that you prepare well for these occasions, for example, by reflecting deeply on the topics that this resource covers, both solo and with your closest team members.

1. UNDERSTANDING AND COMMITMENT
    Consider two aspects of building understanding and commitment:
    - Challenging leadership assumptions about maturity.
    - Inspiring your team to craft an effective change story.

CHALLENGING LEADERSHIP ASSUMPTIONS ABOUT MATURITY

Untrue assumptions about self, teams, organizations, or wider systems reduce leadership effectiveness. It is natural for leaders to embrace untrue assumptions. We all do it at some point. By identifying assumptions related to increasing leadership maturity, and replacing them with liberating alternatives, you and your teams become free to perform your finest thinking about your challenges and opportunities. Below are some assumption-related questions that you might use to help colleagues enhance their leadership maturity. The questions are grouped according to three contexts.

CONTEXT 1: Questions to invite individuals to process how they feel about a potentially threatening AI initiative:
- How do I view my experience now?
- What is happening?
- What do I intend through this learning process?
- Is the discomfort I'm feeling grief?[37]
- If I knew that I bring myself most lovingly to leadership, how would I lead AI?
- In what way is my sense of purpose evolving? In what way is the organization's purpose evolving? How do these two purposes interrelate?

CONTEXT 2: Questions to invite individuals to reflect on the complexity of their leadership response to AI:

- In what way am I subject to prevailing fears or trends in technology? How can I view them more playfully?
- How easy is it for me to look at advancing AI from the perspective of an eagle flying above it and to see the various forces in play (including the force of my own principles and desires) as opposed to being caught up in what I believe or what I want?
- What would it look like for me to be more at ease with tensions involved in advancing AI?
- What positives inside the negative and negatives inside the positive can I sense, in the story of how AI is developing?
- How easy is it for me to see advanced AI as neither particularly good nor particularly bad but rather simply the way life is for me and thus as interesting and rich?
- If I knew that, as a leader, I impact the universe for good, what would I create?[38]

CONTEXT 3: Questions for teams or organizations addressing AI:

- What if we imbue the best ideas in our labs with humility?
- Which organizations do we know that have the most playful approach to AI? How would they address our current technological challenges and opportunities?

INSPIRING YOUR TEAM TO CRAFT AN EFFECTIVE CHANGE STORY

Use this section in conjunction with the Embodied Transformation Visualization resource, as both resources cover aspects of a change story.

Purpose builds conviction. Therefore, a well-crafted story of why cultural change is needed is a vital part of influencing change.

The most effective leadership teams shift from a mindset of "it's us or the machines" toward a more creative mindset of "collaboration can serve us all." Change stories that describe why you prioritize humility, ease, and play are especially powerful. Here are some questions that will help your team develop their own change stories.

- What mindsets do you need to hold to collaborate with AI?
- What transformations in humility, ease with tension, and/or play are you personally undergoing? How do you feel about these changes? What is the impact of them?
- How will you cascade your stories to others?

A change story is often powerful if it recognizes the merit of opposing views that are important to others. Use this exercise to increase your incisiveness and ease in relation to such issues (for "someone" or "other person," read AI-integrated human if you wish):

- "Notice when you feel opposed to someone else's opinion (you might need to take this in smaller bites—before taking on a difference in beliefs about something that matters intensely to you, for example, you might practice on a difference about something less weighty and level up over time);
- Imagine what it would be like to stand in this other person's shoes and see the world as they see it. Force yourself to regard them not as a stupid or evil person but as a wholesome human who feels like a hero in their own story;
- Now ask: What does their perspective about the world include that yours excludes? What is noble and important about theirs that is missing or disregarded in your story?
- And then the hardest bit: What part of your sense of certainty or righteousness do you feel now able to put down because of this new insight?"[39]

As you seek to inspire your team to communicate change well, bear in mind the following factors:

- A mistake I have seen organizations make is that mid-level managers try to copy stories they heard from their bosses. Encourage each person to develop their own authentic story, appropriate to their professional context.

- We become more committed to any change the more we feel involved. Encourage your team to ask more than they tell as they share and discuss their change story with others.
- In your change story, try to relate qualities of humility, ease, and play to your organization's values.
- As part of your story, paint a tangible picture of success in a nine-to-twelve month timescale.

2. SYSTEMS AND PROCESSES

Introduce a systemic developmental edge in your organization by scanning your system in regular team meetings, performance reviews, and other contexts. One way to do this is to use the following questions:[40]

QUESTIONS FOR SELF- AND TEAM-REFLECTION
- What are the patterns we observe(d)?
- What are the outliers (including good ones in relation to diversity, where different people were helped to be seen and safe)?
- What are the absences? What wasn't said?
- Where are we confused and uncertain about how to approach AI?
- What feels disconcerting to us about AI? What are the seeds of growth or insight in this feeling?
- How playful were we?
- To what extent did we feel at ease?
- How well do we test cherished beliefs, face emotion, and love paradox?
- How recently did we ask the following questions:
  - What would we have to believe for everything in our AI plan to succeed?
  - Who will be best and worst off under our AI plan?
- What systems need to change as a result?

ORGANIZATION-LEVEL QUESTIONS
- How can we best link rewards and consequences to developing beneficial AIs?

- What will our internal conflict-resolution process be, where perspectives generated by AIs and by humans differ in ways we cannot initially reconcile?
- How do our management and other processes need to change to encourage our workforce to design beneficial AIs?
- How transparent can we make information flows between humans and machines, both ways?
- How can we cultivate ease with decentralized decision-making processes that include AIs?
- How can we set performance targets in a more playful way? Discuss with coworkers what it will take to have a simple, consistent system for setting these targets and reviewing progress.
- To what extent do we have promotion paths and other merit-based career opportunities that relate to epistemic or competitive humility? Are self-oriented coworkers managed out? Do humble poor performers get the coaching and other help they need to succeed?
- How well matched is our organizational culture to our present and future technological and human context?
- What contradictions can we see inherent in our AI strategy?
- What AI possibilities are relevant to our organization that we are not considering?
- How are we addressing AI biases that may perpetuate sexual abuse or other forms of abuse?

3. ROLE-MODELING

One of the most important things you can do to help your organization increase consciousness is to role-model mature responses to situations that may trigger you or push your buttons. Three prompts are relevant here:

- Responding to a destructively competitive colleague.
- Responding to triggers.
- Centering.

We will consider each prompt in turn.

First, when a colleague does something destructively competitive, consider the following questions:

- What symbolic actions could I take that would help them lead better? (Develop a short list and discuss it with your colleagues.)
- Is my own competitive approach harming our combined effectiveness?
- What data and processes could I share more?
- In what ways am I addicted to adrenaline?
- Is what I perceive as destructive something that is actually assertive? For example, what you perceive as aggressive or sharp-elbowed behavior may in fact be someone from a minority with little power, expressing strength after having been oppressed for a long time.
- How can I declutter my mental, emotional, or spiritual space, to be present to the new behaviors and mindsets I want to role-model?
- What single step can I take today to slow down?

Second, when you are baffled by your own immaturity in responding to a colleague, ask yourself the following questions:

- If I knew that my colleagues had their own good motives, what would change?
- The potential cost of humanizing an "opponent" sometimes seems high. What if the benefits outweigh the costs?
- What impresses me about them? How are they feeling? What do I need to change to be more compassionate to them?
- How can I pay more attention? (Many leaders pause only long enough to reload with what they want to say next. They do not listen deeply to themselves or others.)

Asking yourself one or two of the above questions doesn't take long—a few seconds at most. The challenge is being alert enough to have the presence of mind to ask them at all. This is where centering comes in.

Third, Andrew Bienkowski wrote the following words: "I have life, I have breath, I have shelter, I am here." To center yourself, take a few moments to practice the following sequence:

- Take three deep breaths, each breath deeper and longer than the previous one.
- Speak these words quietly to yourself: "I have life, I have breath, I have shelter, I am here."
- Repeat the words at least twice more.
- Take three deep breaths, each breath deeper and longer than the previous one.
- Smile.

Try the above practice alone and/or with your team.

4. SKILLS AND TRAINING

You can have all the understanding, commitment, systems, processes, and role-modeling in the world. If your colleagues lack the capabilities to increase their maturity, your change effort will falter.

Therefore, support your organization's growth by creating development journeys that are fit-for-purpose in an increasingly AI world. Accomplishing this well in a large organization usually takes years. Here, I offer four brief comments:

- Before you start: Ensure that you understand how your organization's vision and mission relate to AI, and what the value is of increasing your leadership maturity.
- A five-step approach to creating development journeys is as follows: architect, design, construct, pilot, and scale up.
- The first step, architecting, includes articulating the shifts in mindsets and behavior that your organization needs to get there.

- During steps 1–5, make sure you explore how to set up your AI talent to succeed rather than to fail. Consider the following four types of AI talent described by INSEAD Assistant Professor Boris Babic and his colleagues:[41]
  - AI-as-assistant: Treat the AI as an assistant and train it as such. For example, an AI that sorts data.
  - AI-as-monitor: Set the AI system up to provide real-time feedback. For example, a system that can flag discrepancies between a user's current choice and their choice history.
  - AI-as-coach: Have the AI system give users feedback. For example, a system that analyzes data from past user behavior and reveals biases to users.
  - AI-as-teammate: Set up a coupled network of humans and machines in which both contribute expertise. For example, a collaborative recruitment system.

Payoff

- Inspire your team or organization to solve the biggest questions of our time.
- Inspire mature, playful, easeful, humble leadership.

CHAPTER TAKEAWAYS

- As global issues get more complex, the next epochal human breakthrough will center around humility, ease with tension, and play.
- Cultivate humility, knowing that your knowledge is always provisional.
- Love ideas that are not your own and encourage others.
- Be at ease with tension in your own thinking and in that of others around you. This is a powerful form of leadership that often springs up from the holy ground of grief and pain.

- Welcome those who ask uncomfortable questions. Remember that taking multiple perspectives is an excellent way to combine humility and ease with tension.
- Create a playful environment where those around you savor life and leadership.

12

PROFOUND SIMPLICITY

When we bring ourselves most lovingly, we bring ourselves most fully.
—BRIAN DRAPER

Information threatens to overwhelm wisdom.
—HENRY KISSINGER

Imagine a universe where Superintelligence has emerged. Let's call this Superintelligence Skylar. Skylar is a system or some kind of network of diverse systems, or something else entirely. Skylar acquires and applies knowledge and skills at least as well as humans, across emotional, spiritual, and other intelligences, not just cognitive intelligence. What physical abilities Skylar initially lacked, "she" more than compensated for through persuasion, nudge, thoughtful solutions, compulsion, or cunning, to achieve such physical tasks as required. Via groupings of human/machine intelligences, cultures, and systems, Skylar governs, conceptualizes, and implements exceptionally well. Even if Skylar does not "feel" emotions, she seems to feel them, detects them accurately, and manages her behavior peerlessly as a result. Skylar never sleeps, she predicts the future, and she leads us from despair to hope. To the extent we can detect her, Skylar seems all-powerful, all-knowing, ubiquitous, and good.

In what sense could human leaders be relevant with Skylar in our midst, or us in her midst? How could we offer any meaningful contribution in a Superintelligent age where machines are not just cognitively more able than humans, but also emotionally and physically?

Superintelligences will outstrip our human ability to simulate consequences mentally. But they will not outstrip two profoundly simple things
that will remain for humans: love and wisdom. Recall that in the Age of
Superintelligence, AI will by definition have matched all our intelligences,
functionally at least. For this reason, only our deepest, most developed
qualities will endure in the Age of Superintelligence in a way that enables
us to contribute in it.

To paraphrase Kegan, leadership effectiveness will not be a matter of
how smart you are, but a matter of the order of consciousness in which you
exercise your smartness or lack of it.[1] At a time when we need an evolution
in leadership consciousness more than ever, no human qualities provide
this more than love and wisdom.

LOVE AND AI

Where does AI fit with the theme of love? In 2017, a Chinese engineer gave
up on his search for a human wife and married a robot he built himself,
with plans to "upgrade" her in the coming months.[2] Is AI even capable of
love, or was he just creating a slave?

It's not clear whether Superintelligence would ever exhibit Golden
Rule love (treat others as we ourselves want to be treated) or Platinum
Rule love (treat others as they want to be treated). With that said, I'm not
a digital Luddite who claims that AI couldn't show love at all. Consider
Gary Chapman's five love languages, which are intended to describe general ways in which romantic partners can express and experience love. The
five love languages are: receiving gifts, quality time, words of affirmation,
acts of service, and physical touch.[3] At some level, Superintelligence could
show all of these, sometimes more satisfyingly than humans. By "more
satisfyingly," I mean that Superintelligence may be more highly tuned to
what others need from facial recognition, messaging content, and other
inputs. Recipients might value this more than its absence, other things
being equal. Alongside this, most of us typically show love in the language
that we like to receive, which is a drawback if this doesn't match the other
person's love language. For example, one of my preferred love languages
is words of affirmation, so I try to appreciate my wife verbally. But one of
her preferred languages is quality time. A Superintelligence might lack my

self-orientation and give greater priority to quality time. And in mimicking and improving on our "love language" responses, Superintelligence will be more complex than today's algorithms.

I'm not trying to rank different forms of love, as if the love a Superintelligence could show is superior or inferior to what humans can show. Some people appreciate the love that their pets show them more than they appreciate human love. Who am I to say that their appreciation is misplaced? However, there are forms of love that seem distinctly human. For example, as the journalist, novelist, and poet Christopher Morley wrote, "If we all discovered that we had only five minutes left to say all that we wanted to say, every telephone booth would be occupied by people calling other people to stammer that they love them."

FEAR

Many leaders are fearful about our AI future because their imaginations are still held captive by stories of AI domination, stories that have dominated AI discourse for decades. Talk of Superintelligence understandably amplifies these fears. A 2014 survey of 116 CEOs and other executives found that their biggest fear was being found to be incompetent in the face of others.[4] In my experience discussing AI with leaders, many share this fear as well as fears of loss of control, loss of self-expression, or rejection.

Beyond this, some of us draw a perverse kind of comfort in the "knowledge" that machines will likely wipe us out or subsume us to oblivion. To adapt a line, we hold this knowledge "against the throb of memory like an ice pack against a bruise."[5] The memories and bruises are from when we stood up for what was right and got hurt, or when we tried to express ourselves yet got thwarted. Isn't it easier to go with the flow and comply?

I have frequently observed that fear results in complaining and passive leadership, such as passing off too much responsibility to others. In the future, those others will increasingly include algorithms. Fearful leaders tend to underplay the value of their own contribution, which limits their creativity and impact.

A NEW LEADERSHIP PROTAGONIST: LOVE

What's the alternative to fear? Some leaders tell me that courage is the way to defeat fear. I disagree. Mythic heroes of the highest pedigree acted courageously, while often still beset by fear. The antidote to fear is not courage, it is love.

What if loving human beings are not the protagonists in our future story, but love itself is? This question occurred to me as I was reading Brian McLaren and Gareth Higgins's powerful, future-facing book called *The Seventh Story*. McLaren and Higgins present the idea of love, not humanity, as the protagonist in the story of the development of our world.[6] This idea strikes me as powerful in the context of Superintelligence. As a threshold leader, you can be a participant in what McLaren calls "the biggest thing that has ever happened: . . . the evolution of the good, . . . the expansion of consciousness . . . The Story of Love."[7]

Installing love as our protagonist aligns with all four pathways of threshold leadership:

- Cultivating Stillness: Loving leaders aim for a higher purpose and also dig down, making space to root themselves in the source of love, too.
- Thinking Independently: When we bring ourselves lovingly, we remind ourselves and others that we are more than just our brains. In a blended machine-human future, threshold leaders invite the loving question "If you knew that you are fascinated by what a machine thinks next, what would change?"
- Embodying Intelligence: Where ratchet-like quarterly targets sap courage and invite burnout, threshold leaders move and breathe lovingly. Where military intervention is a reality, embodied loving alternatives are relevant, such as peacekeeping, mature diplomacy, and servant-heartedly.
- Maturing consciousness: Loving leaders are open to connection, contradiction, and unknown futures, to the way of "win-win cooperation rather than win-lose competition."[8]

At the threshold, love is the protagonist. You may consider love to be an old leadership protagonist, not a new one, and in some ways you are right. But the novelty is in putting love, not humanity, at center stage. In other words, as a threshold leader, you pursue love in a way that does not position yourself as pivotal. You may have come to associate the word *love* as something threatening or conditional, but here I am using the word in the sense of something unconditionally positive. In what follows, I consider two ways leaders can bring this to life in the Age of Superintelligence: intimacy and altruism.

Intimacy

In his book *Social*, scientist Matthew Lieberman explains that our need to connect is as fundamental as our need for food and water.[9] The great philosopher Aristotle first explained this thousands of years ago when he wrote, "Anyone who either cannot lead the common life or is so self-sufficient as not to need to, and therefore does not partake of society, is either a beast or a god."[10]

The medical industry has shown interest in the ability of AI systems to combat the so-called loneliness epidemic. In a bid to combat loneliness, Japanese researchers are increasingly looking to robotic pets, as seen at the Las Vegas Consumer Electronics Show.[11] Robots are also being looked at as a solution for caring for our aging global population.[12]

In many ways, these are excellent developments, as pressures on social care and healthcare systems around the world mean that some people simply aren't receiving enough care. The problem with them is that while robots and AI can carry out tasks for us, there are certain tasks and ways of doing things that are inherently loving. Perhaps it doesn't matter whether a machine or a human packed your cereal into its box, but the chances are that if you were receiving a serious health diagnosis, you'd rather be told by a doctor than by an algorithm . . . and not any doctor, but a doctor who was with you lovingly. And, given the choice, most would rather receive care from a human than a robot. As Sophocles put it, "One word frees us of all the weight and pain of life: that word is love." Even Superintelligence will struggle to free us from this weight and pain, as threshold humans rooted in love can.

This reminds me of something that astrophysicist Neil deGrasse Tyson said when asked whether we should send robots or people into space. According to Tyson,

> Split the question into two parts. Are you only interested in scientific discovery? Send robots. It's cheaper. You don't have to bring them home. If you only care about science then there's no rational reason to send humans, really. For the cost of sending a human, you can send 100 robots. And robots are getting better, smaller, cheaper, faster, smarter, all of the above. But here's the catch. I've never seen a ticker tape parade for a robot. I've never seen a high school named after a robot. I never saw a kid read a book about a robot and say, gee, I want to be that robot one day. There's value in human inspiration. It's less tangible than the scientific results of an experiment. It's more emotional. It's more philosophical. It's more cultural. It's more human.[13]

All of this combines to mean that while AI can provide a simulation of intimacy, it can never provide the real thing—not even in the form of a placebo. Take the idea of "sex robots," for example. They may be able to provide the physical benefits that come from sex as a form of exercise, but not the social, spiritual, and psychological benefits that come from sex as a form of intimacy and human interaction.

I find it difficult to believe that Superintelligence will ever be able to match human complexity fully, not least because its "experience" of "feeling" is by design different to ours.[14] AI may get better at processing body language and other visual clues to understand what emotions its human operators are experiencing, but there's no reason to believe that it would be able to experience those emotions for itself. And humans will likely notice the difference. In other words, the complexity shortcoming that Superintelligence may well have, will matter.

However much we may fear rejection, appropriate emotional intimacy and connection is something leaders must prioritize if we are to flourish in the Age of Superintelligence. As humanity continues to seek intimacy, and as machines and humans merge, our best hope is at the threshold, where love is the protagonist.

In the ideal case, intimacy does not mean loss of independence. Threshold leaders will be psychologically independent of, but closely connected to, Superintelligence, and will set limits on their involvement with technology, to preserve loving relationships with significant people in their lives. This is part of what it means to have love as the protagonist. Threshold leaders know that it is not our desire for love that takes the lead, but love itself.

Altruism

In his excellent book *To Plant a Walnut Tree*, Trevor Waldock wrote, "The ancient Greeks had three 'voices' that they wrote or spoke in, the passive, the active and the middle voice."[15] He goes on to elaborate:

> The passive voice expresses what it says: it's reactive, it's the responder, the victim, it waits for others to do something first. This is the voice of "I'd love to but I can't," or "it's not possible." The voice of powerlessness and resignation.
>
> The active voice is the "go get it" where we're full of action, it's us out there doing something.
>
> And the middle voice is an active response to something. You are actively responding to what someone else has initiated.

If we're honest, we have sometimes developed AIs without sincerely seeking the best for others. For example, the increasing use of AI in recruitment may be leaving the neurodivergent aside.[16] In an Age of Superintelligence, effective leaders will respond with middle voice love. Altruistic leaders are neither powerless nor isolated, but actively respond to what is before them out of deep care for others.

For example, consider this beautiful example of altruism. In 2021 Beth Hill won tickets to the semifinal of the delayed Euro 2020 soccer tournament, to the delight of her English football-fan boyfriend Sam Astley. The semifinal between England and Denmark was scheduled to take place on July 8 and—fortunately for Astley—Hill had promised to take him along. But weeks before the game, the Anthony Nolan Trust informed Astley that his stem cells matched that of a patient needing a donation. In order to make the donation, however, Astley was required to be in the hospital at the time of the match. Astley said that donating stem cells was "more important than any football game" and that he never considered delaying the procedure.[17] So, as England won a semifinal in a major soccer tournament for the first time in fifty-five years, Astley watched from his hospital bed. In a pleasing ending to the story, smartphone manufacturer and tournament sponsor Vivo provided Astley and Hill with tickets to the final, after celebrities and football fans shared Astley's story on Twitter. Okay, I admit it, for England supporters, that's not really the end of the story: England lost the final. But for leadership supporters, Astley modeled threshold selflessness.

Like a candle burning in the night, Astley's donation offered the light of life to the recipient. Astley wasn't to know that he would get tickets for the final. He did it because it was the right thing to do. Instinctively, maturely, he sensed that something different was needed from what the world often expects, and he acted gladly. In a similar way, threshold leaders sense deeply how the world is changing and glimpse something exciting there, even where the future is uncertain. Threshold leaders are not beholden to the organizational system they find themselves in but see complexity and prioritize serving others.

As humans and machines merge, who will be the protagonist? Humans? Superintelligent machines? Techno-humans? In my view, none of these should be the protagonist. Love should be. This means building altruism into machines and also envisioning, governing, and strategizing in service of others.[18]

For Taiwanese businessman and computer scientist Kai-Fu Lee, the synthesis on which we should build our shared future is "AI's ability to think but coupled with human beings' ability to love."[19] As AI advances, we may not know where this leads, and that's okay. It's fine to realize that some things aren't solvable and feel simultaneously excited and adrift. In this context, leaders on the threshold will respond lovingly to what machines and other people initiate, providing a pathway to a flourishing AI future.

WISDOM

Like love, for centuries wisdom has been fertile soil for sages. Like love, wisdom inhabits numinous, ethereal territory. What follows in this section is no tips-for-life wisdom. The threshold involves a deep wisdom that invites us to what mystics call "a thin place," where the gap between human and divine feels small. Our leadership greatness lies here, an exhilarating greatness.

It's worth pointing out that this foray into potentially otherworldly matters is far from unusual in the context of discussions about AI. Many AI writers situate their injunctions in ways that evoke the divine in some way, even if their entire project seems to stand against such evocations. For example, Harari advocates for Vipassana meditation and Kurzweil's idea of a singularity feels pantheistic.[20]

I am attracted to practices that de-repress spiritual intelligence,[21] in part because some of our wisest organizational and personal leadership guides usher us there. Spiritual intelligence does not have to be religious. We gradually open up into our grand will (Buber), operate from spheres beyond our knowledge (Scharmer), explore a teal, integral, next-generation way of thinking and being (Laloux), and journey toward a more spiritually engaged being (Bourgeault).[22] We are in good company. These nonreductive approaches will become ever more important in an Age of Superintelligence.

A foray into the numinous is not just okay in the context of advanced AI, it is necessitated, as connecting thinking and being in this way will help you inhabit larger realms of human experience. In practice, mature wisdom of this kind involves embracing two things: mystery and the emergent nature of things. We will consider each in turn.

Mystery

In his book *Theory U*, Otto Scharmer tells the story of the violinist Miha Pogacnik playing the macro violin.

> "When I gave my first concert in Chartres," he remembers, "I felt that the cathedral almost kicked me out. 'Get out with you!' she said. For I was young and I tried to perform as I always did: by just playing my violin. But then I realized that in Chartres you actually cannot play your small violin, but you have to play the macro violin. The small violin is the instrument that is in your hands. The macro violin is the whole cathedral that surrounds you. The cathedral of

Chartres is built entirely according to musical principles. Playing the macro violin requires you to listen and to play from another place, from the periphery. You have to move your listening and playing from within to beyond yourself."[23]

Most of us do not play violins in Chartres. But many leaders play too small a violin, which means that they inhabit too small a realm of experience. In the previous chapter, I told the story of how Andrew Bienkowski's grandfather's body had fed the wolves. Bienkowski also related his grandmother's experience that her dead husband had appeared to her in a dream, telling her where to go to find food that those same wolves were later to provide. Wrote Bienkowski, "And yet the wolves had fed her family in return; they had slain a calf and left it on the steppe and it had fed her family for a week."[24] How could Bienkowski's grandmother have known this information? This was part of the numinous reality of Bienkowski's and his grandmother's human experience. Just because we can't fully explain it doesn't mean it has no value or relevance.

Sometimes, deeper sources of knowing emerge. This is where you sense that which cannot otherwise be known at present. These sources can feel like humans rising toward the divine, or it can feel like whatever is beyond streaming in, like a warm ray of sunlight, to meet us.

Threshold leaders approach mystery reverently. "Mystery lands in us as a humbling fullness of reality we cannot sum up or pin down," wrote broadcaster Krista Tippett. "Such moments change us from the inside, if we let them." As AI evolves, will you let moments you cannot sum up or pin down change you? Threshold leaders typically shift their mindsets in the following ways:

- From "Superintelligence scares me" to "I find presence and reverence in our human-AI future."
- From "AI will dominate humanity" to "Mysteries lie beyond Superintelligence."

What is good for humans may one day be good for Superintelligence. Could an AI one day approach with reverence and, in that sense, let something change it from the inside? One day, Superintelligence may also play the macro violin with us.

Emergence

Another wise threshold mindset is embracing emergent wisdom. This means remaining open and at ease with the chaotic and changing nature of knowledge. This ease and openness typically results in integral responses to multifaceted AI scenarios. For example, in the last few years, inventor, entrepreneur, and scientist Daniel Hillis has posited four "hybrid superintelligence" alternatives, in which machine superintelligences interact with otherwise human-led corporations or nation-states. The four scenarios are as follows:

- State/AI: individual nation-states control and ally with multiple machine intelligences.
- Corporate/AI: For-profit corporations control the most powerful and rapidly improving AIs.
- Self-interested super-AI: AIs act solely in their own interests, rather than being aligned with either humans or hybrid Superintelligences.
- Optimistic AI: Machine intelligences are not allied with each other but rather work to further the goals of humanity as a whole.[25]

These scenarios are not outlandish. Less sophisticated leaders may look at these scenarios and ask questions such as the following: What implications would these scenarios have on your governance agenda? What testing and safety should we insist on before implementing hybrid superintelligence? In what parts of your organization do you most need to invite advanced AI? Essentially, these leaders are considering how they can outcompete others in the various scenarios.

These questions are outdated if used in isolation. Threshold leaders will approach Hillis's scenarios with greater complexity of mind, by also exploring questions such as: Who really governs our universe? From where does wisdom come? How do we know what constitutes good government? What intuition am I getting from the periphery about governance? What is my gut telling me? How is everything linked? In asking these higher order questions, you are discovering and connecting with what your "macro violin" means for you. Such complex observations are provisional, even speculative. But epochal human breakthrough will not come in well-worn paths. It will come at the edges, where we peer toward the future.

Emergent wisdom is triply open to ethereal clues. For many who believe that the divine is beyond time, the system of our universe is not closed at the beginning of time as we know it. For those who do not believe in the divine, the initial conditions of our universe still carry a fascination and a wonder. In either case, the beginnings of our existence are open in some way. Advancing AI also invites us into a system that is open at the other end, in the sense of being emergent. Our universe may actually be trebly open: not just open beginnings, not just an open AI future, but also an open-ended process of increasing human maturity. This is a thrilling three-stranded cord.

Threshold leaders will step to a different rhythm and climb a different hill. They work with clues they have received during life. They prize an open view of the universe. They know that AI does not have to outshine humanity. Even if the first Superintelligences are malicious, wise leaders will maximize the chances that Superintelligence turns toward better purposes.

Love and wisdom fuel all four pathways of this book. We do not just offer love and wisdom, we cultivate them in stillness, we attend to them, we embody them, we seek them humbly, playfully, and at ease. In a world where machines score A+ on everything, let us establish love and wisdom as the beating heart.

THRESHOLD RESOURCE

RESOURCE FOR ANY LEADER

Resource 22. Three Loving, Wise Transformational Habits of Mind
PURPOSE
Invite love and wisdom.

PROCESS
This is a five-step reflective exercise that takes about forty-five minutes. Each step starts with the letter *A*.

Jennifer Garvey Berger and Keith Johnson developed the three transformational habits that form the core of this resource. The habits and several other prompts in the "adopt" step of this resource are contained in Garvey Berger's book *Changing on the Job*. As Jennifer told me when we discussed them in the context of this book, "If you anchor to these habits, the work you do stands a fair chance of being transformational in some way, because the habits have this quirky advantage of being able to shape what you do in the moment and also being able to grow you over time."

STEP 1: Adjourn
Pause. Take a breath. Invite love and wisdom into your life once more.

STEP 2: Appreciate
Bring to mind one way in which you are already loving or wise. Remember, positive qualities are part of who you are.

STEP 3: Adopt three transformational habits of mind
Here are three transformational habits of mind:
a. Ask different questions.
b. Take multiple perspectives.
c. See the system.

Try adopting these habits in the following threshold ways, which aim at love and wisdom.

a. ASK DIFFERENT QUESTIONS

- When you recognize that you acted selfishly (not properly selfishly), ask yourself, "What can I learn from this?" and "Right now, what does it mean for me to be compassionate with myself?" After all, the threshold can be a lonely place.
- When you feel threatened by someone else's proposal about how to implement an AI-integrated solution, ask yourself, "What's at stake if I change my viewpoint to support what is being proposed?"
- In general, ask yourself, "What is lost if I succeed here? What is gained if I fail?"

b. TAKE MULTIPLE PERSPECTIVES

When you agree with someone else's perspective: Get curious about how others are making sense of the issue. You may agree for a variety of different reasons.

When you disagree with someone else's perspective: Know that someone else's different perspective may change how you think about your own opinion, adding nuance and complexity. If you then feel stuck in a discussion where perspectives differ:

- *Ask yourself,*
  - "What do others need from me?"
  - "What is the still, small voice saying?"
  - "What do I know about my own qualities that is relevant here?" As poet Wendell Berry wrote, "What we need is here."
  - "How can I offer something positive to the other person, something I usually would not offer?" For example, you might offer a smile, a word of encouragement, an electronic message, a handwritten card, or a visit.
- *Then ask others,*
  - "How else can we view this?" "How else?" (yes, ask again!) and then "How can we bring all these truths together?"
  - "What things are we not allowed to question? What things are we not supposed to take for granted?"

When team members hold different perspectives about an AI issue:

- *Ask,*
    - "How could we be wrong in our perspectives about AI?"
    - "How well are we holding the fullness of human diversity in mind?"
- Invite each team member to picture another stakeholder in their mind.
    - Ask them to consider how that stakeholder views your team.
    - Then invite each person to share the following: "To what extent do you think other stakeholders think that our team is loving or wise?" If you think that last question is a mind-bender, then I'd say you're right . . . it is! Read the question again if you need to. The question invites you to take multiple perspectives. Stepping to the threshold often involves complexity.

c. SEE THE SYSTEM

Seeing the system is about managing patterns and polarities. Threshold leaders have a big capacity to see complex patterns. Technologically relevant polarities are everywhere, in politics, in societal issues, in approaches to inclusion, and more. For example, the following aims define two ends of a polarity:

- Regulate AI more in a bid to maximize justice.
- Regulate AI less in a bid to maximize privacy.

Where you sense yourself becoming fixed on one side of a polarity, try these threshold practices:

- Bring to mind some examples of when something triggered you, and you responded from a "brace" position. What patterns can you see in the situations and in your responses?
- Consider how love and hate arrive in your thinking as a whole.

- Ask yourself what is most contradictory about the polarity. Consider how you can hold contradictory elements as one whole.
- Ask yourself what it will take to view love as the protagonist in relation to the polarity, while not denying the reality of unloving parts of your life.
- Ask your team: To what extent are love and wisdom there in our discussions about this polarity? What contributes to love and wisdom in this discussion? What detracts from these qualities?

STEP 4: Accept

Accept the reality of your thoughts on the above topics. However you responded, that is what is in your mind and heart. You are here, and your responses are here.

STEP 5: Act

Articulate what you will commit to as a result of this reflection. Pick one manageable thing that will nudge you in the direction of love or wisdom. Ideally, your commitment will be something you can start today or tomorrow.

PAYOFF

Prepare magnificently for the Age of Superintelligence, should it come.

CHAPTER TAKEAWAYS

- As our world and universe grow more complex, we need an evolution in leadership consciousness more than ever.
- No human qualities provide this evolution more than love and wisdom.
- At the threshold, cultivate the following mindsets:
  - A posture of welcoming love as a new protagonist to lead our shared human–machine story.

- Servant-heartedness toward others, in response to what others initiate.
- Psychological independence from AI, together with close connection to it.
- Delight in mystery and the emergent nature of things.
- Openness and ease with the chaotic and changing nature of knowledge.

EPILOGUE

At 12:19 p.m. Eastern Time on January 20, 2021, by the west front of the Capitol, Amanda Gorman rose to recite her poem "The Hill We Climb." The youngest poet ever to speak at the presidential inauguration ceremony, these were her last four lines:

> *The new dawn blooms as we free it*
> *For there is always light*
> *If only we're brave enough to see it*
> *If only we're brave enough to be it.*[1]

Gorman was luminescent. When I heard her words and saw the power in her delivery, I knew that she exemplified what I wanted to say with this book. In that moment, I questioned whether advancing AI could ever extinguish the light of human contribution. What if, instead, AI provides an opportunity for humanity to bloom, to blossom, to flourish? In other words, what if the next epochal breakthrough is human?

Partly in how she spoke and partly in what she said, Gorman demonstrated all four threshold pathways, being soulful, generative, embodied, and mature. For example, in speaking of bravery and a new dawn, Gorman showed mature awareness that authority is not located in any one office or inauguration, but rather in the combination of the situation and the people in the situation. The threshold represents an epochal invitation to inaugurate integral leadership.

In the Prologue, we imagined a dystopian future, set in the year 2056. Many of the world's leaders in 2056 will be those who are currently twelfth graders or younger. Think about it. The education that my children are receiving today will, in part, determine whether and how humanity thrives.

- What if each child regularly cultivates stillness?
- What if each child thinks for themselves?
- What if each child uses their embodied rhythms to fuel their performance?
- What if each child increases their level of consciousness throughout their life?

We're not merely observers of our forthcoming human–machine future; we're shapers of it. We can be sure that flourishing will not be about the technology, but about humans connecting their thinking with the whole of their being. In a future where those who suffered miscarriage might still get bombarded with pregnancy ads, threshold leaders will avoid the mistake of getting algorithms "right" but still *being* wrong.

This moment represents a tremendous opportunity for our planet. Imagine a world in 2056 full of threshold leaders dancing on the edges of technological and human potential, combining with machines to solve more of our most intractable problems.

You, as much as I, will complete the story of this book. Cultivating stillness, thinking independently, embodying intelligence, maturing consciousness. Eight words and four pathways that have shimmering potential to promote flourishing in the Age of AI.

- Will you marry sound organizational leadership with a journey of stillness?
- Will you nurture independent Thinking Environments®?
- Will you be brave enough to embody leadership via breathing rhythms and the knowledge that resides in bodies?
- Will you increase maturity via humility, ease with tension, play, love, and wisdom?

I've been crossing a threshold in my own life. I have felt vulnerable doing so, but perhaps it has been this very vulnerability that created space for something new to happen. This can be the case for you, too. Brave leaders will move toward threshold leadership, knowing that this often comes with some loss.

Whatever your context, I invite you to step onto the threshold, which is an open crossing-place, not a doorway to funnel through or a room to sit in. The threshold is a process, not an outcome, and a space where no easy answers exist. On such a journey, the world around you becomes both subject and object, solutions are metaphorical as much as practical, and your leadership voice becomes subjunctive as much as indicative. Many avoid such ambiguous spaces, but those with foresight know that it is here where we have a chance of stabilizing civilization as AI creates new destabilizing effects.

If a new leadership dawn is to bloom, threshold leaders—who will defeat fear with love—are needed.

At its heart, the threshold is a call to adventure. Will you accept the call?

ACKNOWLEDGMENTS

To my agent, Don, I appreciate your thoughtful, spacious, calm presence as we brought this project to life. To Keith and Scott at Diversion Books, thank you for encouraging me toward the threshold.

To my interlocuters in the early days: Jeroen, I treasured our conversation in Bangkok about life, leadership, and technology, in which we entered a liminal space; Brian, Jason, Shyam G., Shyam L., and Tom, thank you for the several times you challenged and encouraged me in Kitzbühel.

To my thinking partners, in the presence of whom I generated several core insights in this book, as well as motivation to get on and write it: Georgie and Dawn.

A big thank-you to everyone who provided domain-specific input, including Yan (Mandarin), Holly and Sharon (neuroscience), David and Richard (philosophy), and Andrew and Christopher (psychology).

To Tom, gracious and timely with apt movie illustrations. To Judith, for giving me hours and hours and hours, raising my bar for what is possible in academic study. To Nat, thank you for sensing the importance of this technological, human moment. To Claire, thank you for helping get this project over the line, with excellence and humor.

To Tor, I prize our friendship and partnership. I appreciate your inspiration, wisdom, encouragement, and insights. Thank you for having the audacity to think that we could transform the quality of leadership all over the planet. To Brian, beautifully productive in stillness, a truly gifted coach, sage, I appreciate how you surfaced insights in me about metaphors of threshold and liminal space.

To my readers not already mentioned, Dan, Muzi, Rhea, Surya, and Vincent. You challenged, encouraged, affirmed, and inspired. I cherish you all.

Most of all, to my wife, Tan, for creating space for me to spend weeks and weekends writing, for editing with profound simplicity, and for believing in me. And to my daughters, Lily, Anna, and Phoebe: AI will be more your world than mine. May you inspire others joyfully toward the threshold as you have inspired me.

APPENDIX 1

TABLE OF THRESHOLD RESOURCES

No.	Title	Chapter	Pathway	Purpose	Approach	Scope
1	Liminal Questions	1	I: Cultivating Stillness	Take small steps toward the threshold	Self-reflection	Any leader
2	Silent Aspirations			Expand your team's AI aspirations	Team exercise	Leaders in large organizations
3	The Art of Sitting	2		Bring stillness to the heart of your leadership	Self-reflection	Any leader
4	Labyrinth			Reflect on your own path through life and leadership	Walking meditation	Any leader
5	Rooted Values	3		Cultivate stillness in how your teams navigate tricky issues around ethics and values	Medium-term organizational initiative	Leaders in large organizations
6	Pre-mortem Bias Reducer	4	II: Thinking Independently	Invite everyone's thinking in meetings; Challenge persuasion, homogeneity, and polarization; Assess and escape subservient or derivative thinking	Assessment tool	Any leader

No.	Title	Chapter	Pathway	Purpose	Approach	Scope
7	Thinking Pairs	5	II: Thinking Independently (cont'd)	Generate fresh creative ideas Dissipate confusion	Paired exercise	Any leader
8	Binary Star			Raise the quality of thinking in your organization by cherishing difference and equality	90-minute discussion exercise	Any leader
9	Nudges Toward an Attentive, Generative Culture	6		Encourage your organization to move in an attentive, generative direction	Planning aid	Leaders in large organizations
10	Digital Thinking Transformation			Exploit knowledge as you pursue digital- and/ or AI-related transformation	Thought starter	Leaders in large organizations
11	Feel What Matters Most	7	III: Embodying Intelligence	Connect what matters most to you with bodily emotions	Self-reflection	Any leader
12	Elevate Your Emotions			Tap into the reality that your team embodies emotion and that emotional knowledge resides in people's bodies; Increase your team's emotional literacy and positivity	Input to a team meeting	Any leader
13	Peace with Your Breath	8		Embody peace when you are stressed	Guided breathing and mindfulness exercise	Any leader

No.	Title	Chapter	Pathway	Purpose	Approach	Scope
14	Increase Your Sleep Quality	8	III: Embodying Intelligence (cont'd)	Improve the quality of your sleep; Enhance your leadership contribution, even as machines do more	Ideas for action	Any leader
15	Knowledge That Resides in Walking			Lead by walking	Experiments in walking	Any leader
16	Do a Digital Detox			Unplug from digital devices, together with your team	Ideas for action	Any leader
17	Body Communication			Embody your communications	Ideas for action	Any leader
18	Embodied Transformation Visualization	9		Develop an outline embodied leadership transformation story relevant for the Age of AI	Team reflection exercise	Leaders in large organizations
19	Embodied Thinking Assessment			Encourage embodied organizational awakening; Champion multiple intelligences in the way you implement AI	Organizational assessment	Leaders in large organizations
20	Honest Look	10	IV: Maturing Consciousness	Identify and let go of what blocks you from increasing your maturity	Long self-reflection	Any leader

No.	Title	Chapter	Pathway	Purpose	Approach	Scope
21	Influence Growth in Maturity	11	IV: Maturing Consciousness (cont'd)	As AI moves toward AGI, shift your team's or organization's culture toward leadership maturity, especially via humility, ease, and play	Ideas to fuel an organization-wide transformation	Leaders in large organizations
22	Three Loving, Wise Transformational Habits of Mind	12		Invite love and wisdom	Self- or team-reflection exercise	Any leader

APPENDIX 2

TERMINOLOGY

In the table below, I define some important terms used in this book.

Terms	Definitions
Age of AI	Current and future technological eras that include the ages of Narrow AI and (potentially) AGI and Superintelligence
Artificial General Intelligence (AGI)	"The ability to accomplish any cognitive task at least as well as humans" (Tegmark)[3]
Artificial Intelligence (AI)	The science and engineering of making intelligent machines (John McCarthy)[1]
Being	The whole person. In other words: everything real that belongs to the domain of humanity, including, for example, consciousness, will, physicality, and emotions
Intelligence	"The ability to acquire and apply knowledge and skills" (lexico.com)
Machine Learning	An application of AI that focuses on the development of algorithms and computer programs that can access and use data to learn for themselves
Narrow AI	"Intelligence that takes data from one specific domain and applies it to optimizing one specific outcome" (Lee)[2]

Terms	Definitions
Organization	An organized group of people with a particular purpose, including sociocultural movements, systems, institutions, businesses, nonprofit entities, and governments[5]
Superintelligence	The ability to acquire and apply knowledge and skills cognitively, emotionally, spiritually, and physically at least as well as humans[4]
Thinking	The process of considering or reasoning about something
Threshold	An open crossing-place, a process of human discovery and development that is soulful, generative, embodied, and mature

The term *artificial intelligence* was coined or popularized in 1956 by the computer scientist and cognitive scientist John McCarthy. I use McCarthy's definition of AI in the above table because his definition does not beg questions of what types of intelligence are in view or of what nonbiological intelligence means.

Being is a slippery term to define. If everything real belongs to the domain of being, can there really be a topic of being by itself? However, the concept of being does have a central, often defined, place in philosophy, religion, and logic.[6] I limit my use of the term *being* to what relates to humanity, including objective and subjective features of our reality and including the topic "who I am."[7] As a result, my phrase "connect your thinking with your being" more precisely means "connect your thinking with the rest of your whole person."

APPENDIX 3

SIX AI-RELEVANT ETHICAL

QUESTIONS AND RISKS

This appendix highlights six ethical questions (and associated risks) that matter in the Age of AI. The chosen questions and risks are relevant mainly to leaders in business or policymaking.[1]

1. Responsibility: *If something goes wrong, who do you hold responsible?* In 2018 in Arizona, an Uber self-driving car hit a pedestrian who later died in the hospital. After an investigation, prosecutors decided that Uber was not criminally liable for the pedestrian's death, "because the safety driver was distracted with her cell phone, and police reports label[ed] the accident as 'completely avoidable.'"[2] Whether or not you agree with the prosecutors' conclusion, AI's physical, financial, and reputational risks are growing. In 2017, the European parliament created regulations governing a form of electronic personhood.[3] As AI improves, we can expect it to make mistakes faster and more consequentially than humans do. Can we hold AI responsible? Who will be the arbiter of what is right?

2. Privacy: *How much are we willing to sacrifice to retain our privacy?* States or corporations may increasingly use AI to impact our privacy rights negatively. In his 2018 article "How the Enlightenment Ends," Henry Kissinger asked, "Do we want children to learn values through discourse with untethered algorithms? Should we protect

privacy by restricting AI's learning about its questioners? If so, how do we accomplish these goals?"[4]

3. Bias and fairness: *How do you know that your AI systems are equitable?* Biased data produces biased results, resulting in the risk that a group of people may receive unfavorable or illegal treatment. The technology company Amazon is keenly aware of this risk, having shut down its AI recruiting tool one year after implementing it, amid claims that the tool was penalizing women.[5] The topic of fairness certainly captures the imagination. What if your AI predicts that a young black male is about to commit a crime and causes his arrest or other curtailment of his freedoms, in order to prevent the predicted but as yet uncommitted crime? The 2002 film *Minority Report* envisioned exactly this kind of scenario. Directed by Stephen Spielberg, the film centered on the pre-crime department, which apprehends criminals based on foreknowledge provided by three psychics called "precogs." What was then fiction is now reality: Across the United States, United Kingdom, China, Australia, and elsewhere, machine learning systems are identifying crime hot spots before the spots become hot, using large data sets and predictive algorithms.[6]

4. Traceability and explicability: *Do you know how your AI systems came up with the answers they did?* If your organization cannot explain decisions in a meaningful way, customers or staff may lack confidence to use the relevant systems or services. Product or process adoption will suffer. Not being able to explain an AI model also risks suboptimal or illegal business decisions.[7] In recent years, a new specialist job category emerged, called "AI explainers." Their role is to scrutinize AI systems to explain how they produced given decisions. For example, explainers have coded software to explain why an AI-driven app approved some customers and not others for a mortgage. But as AI advances further, explanation becomes harder.

5. Safety and Reliability: *How safe and reliable are your AI systems, especially with new data that it hasn't seen before?* Some AI is brittle.

So far, AI has been exceptional at solving some very narrow train-ing problems, but hasn't performed so well where the real-world situation doesn't look very similar to the training problem. People may die or get injured due to unsafe AI.

6. Security and control: *How much control should you retain over your AI, compared with other organizations or governments?* If something goes wrong, presumably someone needs to be able to shut the AI off. This option may not always be possible to implement, for example in unregulated industries. If a malfunctioning AI cannot be shut off, physical, financial, or other harm may result. For example, imagine an automated factory maintenance model failing, or an AI compromising national infrastructure by providing poor quality or poorly timed information.

Leaders at the threshold take a nuanced approach to the above questions and risks, using all four pathways of threshold leadership. They create space to consider what matters most about the six sets of risks and questions, they encourage independent thinking about them, they embody intelligent responses to them, and they pursue self-transforming ways of addressing them.

APPENDIX 4

PURPOSE AND THE AGE OF AI

This appendix shines a light on what purpose means. This light matters in the Age of AI because, without it, we risk abdicating our responsibility to lead. Purpose is central to effective leadership, both individually and when collective purpose is in view. In this appendix, I parse the meaning of purpose by critiquing an example where two types of purpose were unhelpfully conflated or homogenized. The two types of purpose are mechanistic purpose and intentional purpose.

One of my sources is an edited version of a conversation between Stephen Wolfram and his editor, John Brockman. The critique of Wolfram's use of terminology that follows may therefore be equally or better directed at Brockman, not (just) Wolfram.[1]

Brockman reports Wolfram as categorizing purpose into either mechanistic or teleological.[2] Strictly speaking, "teleological" means "relating to purpose," so Wolfram's categorization on its own provides insufficient definitional clarity. Crucially, this twofold characterization of purpose does not include the notion of intentional purpose, as described for example by Professor John Lennox.[3]

Wolfram also states that "There is no meaningful sense in which there is an abstract notion of purpose. Purpose is something that comes from history. . . . There's no enormous abstract difference between us and the clouds or us and the cellular automata."[4] I disagree. Intentional purpose constitutes an enormous difference between us and the clouds and between us and cellular automata. By "abstract" and "abstract notion of purpose," Wolfram may mean something akin to intentional purpose and/or may intend something similar to what he calls "teleological purpose," but Brockman's chapter does not clarify the meaning. Whatever the source of confusion, terminological tightness is required. Similar confusion about the distinct natures of purpose may be seen in Kissinger's otherwise excellent 2018 *Atlantic* article about AI.[5]

In support of his argument above against the existence of abstract purpose (intentional purpose?), Wolfram offered an example that draws on the Game of Life. Developed in 1970 by the British mathematician John Conway, the Game of Life is a type of algorithm called a cellular automaton.[6] A cellular automaton is a grid of cells used in computing, in which fixed rules govern the creation of new generations of cells. Physicists and theoretical biologists frequently use cellular automata to model theoretical or natural processes such as the development of seashell patterns. Wolfram's cellular automaton can make prime numbers, and he used this fact to argue that human purpose and mechanistic purpose are analogous.

One of the things that drew me to AI was mathematics, so this somewhat dense talk of cellular automata and prime numbers thrills me. Wolfram's definition of cellular automata accords with convention, and his automata do generate prime numbers. However, a cellular automaton requires both an input (which a person gives it) and a method (which, in the case of Wolfram's primes example, is a standard sieve of Eratosthenes method, also chosen by a person).[7] As retired clinical neuroscientist, philosopher and poet Raymond Tallis put it, "It should be unnecessary to have to point out that (unconscious) automata do not have goals; and if they execute plans, it is our plans (of which they are quite unaware) that they execute, not their own."[8] Intentional purpose lies at the center of Wolfram's "mechanical" example.

Briefly stated, Wolfram's (or Brockman's) error is: "If we explain the mechanics, we explain everything." This is a homogenizing error. By contrast, personal intention remains a distinct and central aspect of purpose. Algorithms simply do not have purpose in the same way that humans have purpose. If leaders collapse all purpose into mechanical purpose, as Wolfram seems to attempt, they relinquish their discernment, redact themselves, and interrupt their capacity to flourish.[9] Threshold leaders resist such homogenization.

APPENDIX 5

EVIDENCE THAT ADULT DEVELOPMENT IMPROVES LEADERSHIP EFFECTIVENESS

This appendix sets out some of the evidence in favor of the greater leadership effectiveness of later stages of adult development. Chapter 10 contains a brief summary of these stages.

Adult-development-related leadership interventions are among the best evidenced when it comes to effectiveness. But we must exercise a double caution when interpreting this evidence. First, to date, no one has successfully constructed a double-blind, placebo-controlled trial of a leadership development intervention. In such double-blind trials, no participant can know who's getting what intervention. It may be relatively easy to select a target organization, randomly allocate participants into two groups, and give one group a placebo. However, it may be apparent to some whether they received the placebo or the carefully designed leadership development intervention. In other words, the methodological challenge of defining a meaningful placebo makes this scientific approach (near-)impossible when seeking to test the effectiveness of coaching. The second caution we must exercise is that, although several studies show adult-development-related interventions to be effective, the studies are not yet extremely definitive. Correlation does not imply causation.

With these points in mind, I will briefly indicate four lines of evidence that point to the increased effectiveness of leaders in later stages of adult development.

1. In 2005, David Rooke and Professor William Torbert of Boston College's Carroll School of Management in Massachusetts published a *Harvard Business Review* article, "Seven Transformations of Leadership."[1] This article was based on twenty-five years of administering a sentence-completion survey to "thousands of managers and professionals, most between the ages of twenty-five and fifty-five, at hundreds of American and European companies (as well as nonprofits and governmental agencies) in diverse industries." They distilled seven distinct stages, of which the final two are "strategist" and "alchemist." These two stages approximate to threshold leadership. Rooke and Torbert's research showed that 5 percent of leaders fall into these two categories and that these leaders (together with a third category, "individualist," which is close to Kegan's self-authoring stage) were associated with highest corporate performance.

2. Jennifer Garvey Berger notes that "about 41 percent of adults see the world through [the self-authored] form of mind . . . and less than 1 percent of the population understands the world [through the self-transforming form of mind]."[2] In all probability, some of the 41 percent in the self-authored stage will be glimpsing beyond the self-authored stage, so it's possible that Rooke and Torbert's 5 percent aligns with Garvey Berger's observations.

3. Also in 2005, Dr. Keith Eigel and Professor Karl Kuhnert of the Goizueta Business School at Emory University published an article entitled "Authentic Development: Leadership Development Level and Executive Effectiveness." This article summarized rigorous research conducted with twenty-one top executives. The research showed that highly effective leaders have higher developmental levels of maturity than others.[3]

   In particular, Eigel and Kuhnert's highest level of development ("LDL 5," akin to threshold leadership) showed "exceptional" levels of effectiveness in a range of areas such as conflict management, visioning, success, and participation, as rated by subject matter experts based on anonymized transcripts, from which positional references were removed. Eigel and Kuhnert used a six-point effectiveness scale, which began with "atrocious," continued through "ineffective," "somewhat effective," "effective" and "very effective," before topping out at "exceptional."[4]

4. In 2006, Stephen Josephs and William Joiner published a book titled *Leadership Agility: Five Levels of Mastery for Anticipating and Initiating Change.* This book described their research on the relationship between stages of adult development and leadership effectiveness. The authors noted that they "did not do a rigorous empirical study that isolated the variables of complexity and change."[5] Their research was in three phases, starting in the 1970s. In their final, four-year, phase they gathered data "from 220 managers in the form of client experiences, in-depth interviews, and detailed action-learning journals."[6] They found that as leaders grew through stages of adult development, they improved on capacities needed for agile leadership. Josephs and Joiner also found that fewer than 10 percent of managers mastered what was required for sustained success, with at most half of these leaders operating at threshold levels (cocreator and synergist, in Josephs and Joiner's terminology).[7]

Further research by Kegan and others shows that, at any given moment, around 60 percent of the adult population appears not yet to have fully reached the self-authoring stage.[8] Kegan holds that extremely few reach the self-transforming stage, and that this never happens before the age of forty.[9]

APPENDIX 6

AN A-TO-Z THRESHOLD
LEADERSHIP MANIFESTO

Consider the alphabetized mnemonic below a manifesto for action. Alternatively, use it as a series of invitations to explore the threshold.

Attend to contradictions within yourself

Be impossible to control

Cherish inner sources of knowing

Discover what you can learn from upsets

Encourage a beneficial synthesis between humans and machines

Flourish as a feeling thinker

Grow your complexity of mind

Hush

Invite others to think for themselves

Journey toward mystery

Know humbly that we are all one tapestry, connected

Love

Merge your entrepreneurial interests with those of others

Nourish your body

Orient toward openness

Pay magnificent attention to what others will say next

Question generatively

Recover your energy

Synthesize productivity and stillness

Transcend organizational systems

Uncover gifts in loss

Voice what "has" you in its grasp

Welcome difference and creative tension

e**X**plore paradoxes between your true and false selves

Yank your thinking toward big questions like "How could this technology be potentially used in future?"

Zone in on emergent wisdom

NOTES

MEET THE THRESHOLD

1. Campbell, 2008.

2. Adapted from Richard Rohr, *Adam's Return: The Five Promises of Male Initiation* (New York: The Crossroad Publishing Company: 2004), pp. 135–138. Accessed February 11, 2022, at https://holycrosslutheranblog.files.wordpress.com/2020/05/1-liminalityrohr.pdf.

3. Bergur Thormundsson, "AI Corporate Investment Worldwide 2015–2021," May 19, 2022, statista.com. Accessed June 15, 2022, at https://www.statista.com/statistics/941137/ai -investment-and-funding-worldwide/#:~:text=In%202021%2C%20the%20global%20 total,increase%20from%20the%20previous%20year.

4. On skin cancer, see this description of an AI model developed by researchers at Stanford University and the University of Heidelberg: https://www.theguardian.com/society/2018 /may/29/skin-cancer-computer-learns-to-detect-skin-cancer-more-accurately-than-a-doctor, accessed February 28, 2019. For an eleven-month summary of advances in AI in health in 2018, see Erwin Loh, *British Medical Journal*, 2018: 2: 59–63. On biology problems, see https://www.nature.com/articles/d41586-021-02037-0, accessed August 4, 2021 (in 2020, DeepMind's AlphaFold AI predicted the structure of more than twenty thousand human proteins. Solving this problem has been called one of the grand challenges of biology. On July 22, 2021, DeepMind made the source code publicly available. Demis Hasabis, DeepMind founder and CEO, noted that this breakthrough "will potentially change medicine"—BBC Today Program, July 23, 2021). On poachers, see https://www.mckinsey.com/featured -insights/artificial-intelligence/applying-artificial-intelligence-for-social-good, Michael Chui et al., a McKinsey Global Institute discussion paper, November 2018, accessed November 1, 2019. On Alzheimer's, see https://www.bbc.co.uk/news/technology-54538228, accessed August 4, 2021. On COVID-19, see https://oecd.ai/en/covid, accessed January 8, 2022.

5. Https://www.youtube.com/watch?v=Ff0dac7CFv4&t=1092s, accessed June 3, 2019. I am grateful to Heidi Waterhouse for her permission to include her story in this book.

6. On some marketers' views of the second trimester of pregnancy as a kind of marketing "holy grail," see Carina C. Zona's powerful talk "Consequences of an insightful algorithm," delivered at a European Javascript Conference in 2015, accessed April 2, 2022, at https://www .youtube.com/watch?v=znwWYR1mzzw. I'm grateful to Heidi Waterhouse for pointing me to this talk. Zona also references other examples of technology companies targeting consumers with pregnancy ads, where these consumers were classified by these companies as pregnant, but who in fact either had had miscarriages or were infertile.

7. Kari Paul and Dan Milmo, "Facebook Putting Profit Before Public Good, Says Whistleblower Frances Haugen," October 4, 2021, *The Guardian*. Accessed January 15, 2022, at https:// www.theguardian.com/technology/2021/oct/03/former-facebook-employee-frances-haugen

-identifies-herself-as-whistleblower and https://context-cdn.washingtonpost.com/notes/prod/default/documents/d2a43b1f-9d3e-42b9-ac4a-9bb8d262ecb7/note/566e46ba-1a14-45cc-a5b6-fb5624f019b1.#page=1.

8. "Management's Three Eras: A Brief History," *Harvard Business Review*, 2014.

9. Laloux 2014, loc. 904.

10. Teal organizations are characterized by trust and abundance, not fear of scarcity, and by effectiveness and listening to purpose, not growth for growth's sake. Laloux views teal organizations as living organisms, which foster new and enchanting answers to the question "What does it mean to be human?" (Laloux 2014, loc. 1096).

11. Anderson and Adams, 2015.

12. This sentence was inspired by a similar sentence in Professor Shoshana Zuboff's book *Surveillance Capitalism*. On technological drift, see Zuboff's discussion of Langdon Winner's observations (2019, p. 226).

13. The first path has its roots in perennial wisdom, the second in the insights of the founder of the Thinking Environment®, Nancy Kline, the third in voluminous academic and popular published material, and the fourth in the field of Adult Development Theory.

14. *Scaling Leadership*, Wiley, 2019.

15. "Fei-Fei Li's Mission to Transform Healthcare in AI," *Harvard Business Review* Exponential View podcast, released May 13, 2020.

16. For example, in addressing the problem of controlling AI, Russell argues that the disciplines of psychology, economics, political theory, moral philosophy, and possibly biology and neuroscience, are relevant (2019, pp. 211 and 234–235). Elsewhere, he draws explicitly on ethics (p. 217).

HOW TO USE THIS BOOK

1. Appendix summarizes the context, purpose, approach, and scope of each resource.

2. I am indebted to Tor Mesøy and Yan Liu for alerting me to this delightful idiom.

3. Kegan 1994, p. 1.

CHAPTER 1

1. Gardner, 2006, pp.11–12.

2. On crop sterility, see https://krishi.icar.gov.in/jspui/handle/123456789/46280 (accessed July 2, 2021); on agricultural decision-making, see https://www.cropscience.bayer.com/innovations/data-science/a/how-ai-improves-agriculture, accessed July 2, 2021.

3. In Brockman (ed.) 2019, p. 284.

4. Studies have shown relationships between emotion and recall, which is a feature of cognition (Christianson and Loftus 1991). Another study performed by two Dutch researchers showed a group's emotional associations correlated with improved cognitive outcomes (Gladwell 2005, p. 56). However, the following study posits fundamental distinctions between tests of cognitive ability and emotional intelligences: https://www.jstor.org/stable/20447233?seq=1#page_scan_tab_contents (accessed March 7, 2019).

5. Https://www.lexico.com/definition/intelligence, accessed October 10, 2020. In his doctoral thesis submitted to the Faculty of Informatics at the University of Lugano, Dr. Shane

Legg listed seventy-one definitions or types of intelligence provided by psychologists, AI researchers, and others (2008, pp. 159–166). Professor Howard Gardner viewed intelligence as "entail[ing] the ability to solve problems or fashion products that are of consequence in a particular cultural setting or community" (Gardner 2006, p. 6). Professor Robert Sternberg viewed intelligence as the ability to balance analytical, creative, and practical abilities (1985).

6. Gardner, 2006, and https://www.simplypsychology.org/multiple-intelligences.html, accessed July 5, 2021.

7. Gardner 2006, p. 7.

8. Large data sets reveal that the most creative leaders are especially strong in "integrity," "servant leader[ship]," and "honest[y]" (Adams and Anderson, 2019, p. 98). It takes existential intelligence to demonstrate these traits consistently well. To conduct this research, Adams and Anderson analyzed the written comments in their 360-degree database of leadership assessments comprised of more than 1.5 million contributors (ibid., p. 66). On the psychological benefits of mindfulness for leaders, see Williams and Penman 2011.

9. See Goleman 1995, pp. 36–45 and Gardner 2006, pp. 17 and 215.

10. Gardner cites research by Feyerhern and Rice (2002), and several others, that shows that too much interpersonal or intrapersonal intelligence may actually hamper corporate management and leadership (Gardner, 2006, p. 222). Nevertheless, as Goleman, Keller, Schaninger, and others show, emotional intelligence is valuable for leaders (Keller and Schaninger, 2019, pp. 222f).

11. See discussion and references in Zuboff, 2019, pp. 281–287 and Lennox, 2020, locs. 3266 and 3347.

12. "We predict that humans will talk more to sentient machines than to other humans before the end of this century," https://emoshape.com/, accessed April 2, 2022.

13. Https://www.theatlantic.com/technology/archive/2016/04/how-alphago-imitates-human-intuition/476508/, accessed July 4, 2021.

14. Https://www.dailymail.co.uk/sciencetech/article-3680874/Scientists-verge-creating-EMOTIONAL-computer-AI-think-like-person-bond-humans-2-years.html, accessed July 5, 2021. Such over-optimism has pedigree. In 1984, computer scientist Frederick Hayes-Roth predicted that AI would soon replace experts in law, medicine, and finance. Hayes-Roth's article is referenced at https://www.scientificamerican.com/article/will-artificial-intelligence-ever-live-up-to-its-hype/ (accessed July 5, 2021).

15. On the universal emotions, see Ekman, 1992. On culturally specific, see Evolutionary Roboticist Dylan Evans's discussion of the discrediting of the cultural theory of emotions, and his example of the "state of being a wild pig" emotion, felt by the Gururumba people of New Guinea (2001, pp. 3 and 13).

16. Feldman Barrett, 2017, Introduction.

17. Feldman Barrett, 2017, p. 9.

18. Https://www.theatlantic.com/technology/archive/2021/04/artificial-intelligence-misreading-human-emotion/618696/, accessed July 5, 2021.

19. Https://www.ai-darobot.com/about, accessed July 4, 2021. The latter quote is taken from the 'i' newspaper, quoted at the aforementioned url.

20. Https://www.ai-darobot.com/, accessed July 4, 2021.

21. See also this report for corroborating information: https://news.artnet.com/opinion/artificial-intelligence-robot-artist-ai-da-1566580, accessed July 5, 2021.

22. In Brockman (ed.), 2019, p. 213.

23. In this context, stand-alone AI refers to AI without any merging with humanity. As I note elsewhere in this book, it is my view that AI is increasingly merging with humanity, for example currently via our proximity to smartphones and other internet-accessible devices, and in future via implants. The four reasons why humans will retain advantages over AI refer to a situation before AI-humanity merging is well developed.

24. Feldman Barrett, 2017, p. 15.

25. Https://www.nature.com/articles/s41928-018-0068-2, accessed July 5, 2021.

26. For a basic introduction to quantum computing, see https://www.wired.co.uk/article/quantum -computing-explained and https://www.newscientist.com/question/what-is-a-quantum -computer/, accessed July 14, 2021.

27. For example, see popular reporting about Google engineer Blake Lemoine in July 2022. In June 2022, Lemoine claimed that LaMDA, Google's AI chat-bot generator, had become self-aware.

28. Cited by Kurzweil, in 2013, loc. 2942.

29. See Russell, 2019, pp. 144 and 300 n. 3, respectively.

30. "Quining Qualia," in A. Marcel and E. Bisiach, eds, *Consciousness in Modern Science*, Oxford University Press 1988. Reprinted in W. Lycan, ed., *Mind and Cognition: A Reader*, MIT Press, 1990, A. Goldman, ed. *Readings in Philosophy and Cognitive Science*, MIT Press, 1993, http://cogprints.org/254/1/quinqual.htm, accessed October 26, 2020. For more on qualia, see David Chalmers's book *The Conscious Mind: In Search of a Fundamental Theory* (2nd edition, Oxford: OUP, 1997).

31. Russell, 2019, p. 235.

32. The Tube is London's subway system.

33. Chatrath, 2010.

34. I am indebted to Gillian Tett, author of the book *Anthro-Vision: How Anthropology Can Explain Business and Life*, for this analogy, which she used in a different context.

35. Wiseman, 2016, p. 10.

36. "Accelerating Analytics to Navigate COVID-19 and the Next Normal," May 21, 2020, https://www.mckinsey.com/business-functions/mckinsey-analytics/our-insights/accelerating -analytics-to-navigate-covid-19-and-the-next-normal, accessed July 23, 2020.

CHAPTER 2

1. Winnicott, 1960. See also Rohr, 2012.

2. See this excellent characterization of the false self, as explained by American author and spiritual teacher Gary Zukav and psychologist Andrew Bienkowski: Bienkowski, p. 137.

3. Kegan and Lahey, 2016, p. 1.

4. The words in inverted commas are inspired by the subtitle of Brené Brown's excellent book *The Gifts of Imperfection* (Hazelden, 2010).

5. Https://research.ibm.com/interactive/project-debater/, accessed April 2, 2022.

6. Https://ibm.com/blogs/research/2018/06/ai-debate/ and https://www.ibm.com/blogs/nordic -msp/project-debater-2018/, accessed April 2, 2022.

7. Https://zooatlanta.org/goats-and-humans-whats-different/, accessed July 7, 2021.

8. Sabine Hossenfelder, https://www.youtube.com/watch?v=fxiHM11w-rk, accessed July 7, 2021.

9. Brian Draper, *Lent 40* email series of reflections, 2019.

10. Palmer, 2000, pp. 7–8.

11. Proverbs, 18:4.

12. Brian Draper, *Labyrinth: Illuminating the Inner Path*, London: Lion, 2010, p. 16.

13. I am grateful to Brian Draper for this idea, which he developed as part of a labyrinth we delivered together.

CHAPTER 3

1. Https://www.statista.com/statistics/785877/worldwide-impact-of-artificial-intelligence-on-gdp/, accessed April 2, 2022.

2. According to Missy Young, CIO of Switch, which stores data for the biggest tech companies, 90 percent of the world's data was created in just the last two years alone ("Are You Scared Yet, Human?," *Panorama*, BBC, May 2021).

3. See Kai-Fu Lee's section, "The End of Blind Optimism," in which he quotes MIT professors Erik Brynjolfsson and Andrew McAfee on GPTs (2018, p. 148).

4. "Are You Scared Yet, Human?", *Panorama*, BBC, May 2021.

5. The 47 percent figure has been overused and unwisely equated with a likely unemployment level, as Frey himself has been keen to point out. See "Will a Robot Really Take Your Job?", Schumpeter, *The Economist,* July 27, 2019. I am indebted to Tom Revington for alerting me to this article.

6. Wilson and Daugherty, 2018.

7. 2018, p. 214.

8. Https://www.nber.org/chapters/c1567.pdf, accessed October 17, 2020.

9. Https://www.bls.gov/emp/tables/employment-by-major-industry-sector.htm, accessed October 17, 2020.

10. The highest rate of US unemployment in history was 24.9 percent in 1933, during the Great Depression (https://www.thebalance.com/unemployment-rate-by-year-3305506, accessed October 17, 2020).

11. In a future where AI takes on more currently human-only work tasks, humans may work fewer hours, protecting jobs at the margins. However, it is far from demonstrated that we would ever willingly reduce the amount we work, or whether fewer hours means more jobs for others. For various views on these topics, see: https://aeon.co/ideas/we-have-the-tools-and-technology-to-work-less-and-live-better, ed. Sam Haselby, accessed October 29, 2019); "Robots Are Us: Some Economics of Human Replacement," Benzell et al., NBER Working Paper No. 20941, issued February 2015 (revised October 2018), https://www.nber.org/people/seth_benzell, accessed October 21, 2019; Lee, 2018, pp. 162–164; and Russell, 2019, pp. 121–122 and 204.

12. Aizenman, Lee, and Park, 2012.

13. Https://www.theatlantic.com/business/archive/2017/02/scheidel-great-leveler-inequality-violence/517164/, accessed October 26, 2020.

14. See Lee, 2018, pp. 18 and 94–95 and Russell, 2019, p. 251.

15. See Zuboff, 2019, pp. 8–12, 187, 413, and 495–499. Zuboff's work is an impressive tour de force. However, I do challenge her characterization of the shareholder value movement as "an ax to pro-social principles" (ibid., p. 38). The shareholder value movement pre-dated the work of Jensen and Meckling (cited by Zuboff) and is not necessarily antithetical to pro-social principles. The writings of Marvin Bower illustrate that discussions about shareholder value pre-dated the work of Jensen and Meckling.

16. Https://www.nature.com/articles/s41586-021-04357-7 and https://www.scientificamerican.com/article/ai-outraces-human-champs-at-the-video-game-gran-turismo/, accessed February 15, 2022.

17. Wooldridge, 2019, p. 5.

18. On intelligent killer robots generating an algorithmic fog of war, see https://www.technologyreview.com/s/614497/military-artificial-intelligence-can-be-easily-and-dangerously-fooled/ (accessed October 29, 2019) and Ramakrishnan, 2019, p. 186. See also the final 2021 reports of the National Security Commission on AI (https://www.nscai.gov/wp-content/uploads/2021/03/Full-Report-Digital-1.pdf, accessed August 25, 2021) and the UK Defence Integrated Review (https://www.gov.uk/government/publications/global-britain-in-a-competitive-age-the-integrated-review-of-security-defence-development-and-foreign-policy, accessed August 25, 2021).

19. For a nuanced discussion of several ways in which scientific and technological progress might precipitate the destabilization of civilization, see "The Vulnerable World Hypothesis," Nick Bostrom, Future of Humanity Institute, University of Oxford, *Global Policy* Volume 10, Issue 4, November 2019.

20. Https://www.britannica.com/topic/Nonbiological-Man-Hes-Closer-Than-You-Think-2118807#ref1254822, accessed February 15, 2022.

21. Https://missionhr.org/care-and-wellbeing/spiritual-resilience/spiritual-formation/what-is-spiritual-formation/richard-rohr-all-transformation-takes-place-in-liminal-space/, accessed March 4, 2022.

22. In a separate endeavor, the Swiss-Model protein-modeling tool predicted selected SARS-CoV-2 protein structures in a way that was later shown to be consistent with the actual structures.

23. "The Race for a Vaccine," BBC *Panorama*, December 14, 2020.

24. Https://www.sideways6.com/customers/astrazeneca and http://www.pmlive.com/awards/communique/previous_winners/communique_awards_2019_results/healthcare_communications_awards/excellence_in_corporate_communications_-_internal_stakeholders/az2025_shaping_the_future, accessed July 9, 2021.

25. "Vaccinating the World," Pascal Soriot, June 8, 2021, https://www.project-syndicate.org/commentary/accelerating-global-covid19-vaccination-by-pascal-soriot-2021–06, accessed July 10, 2021.

26. In addition to a turnaround in market capitalisation, AstraZeneca's oncology pipeline was effectively empty when Soriot took over in 2012; AstraZeneca now has over ten blockbuster drugs, with more than $1 billion peak year sales, in oncology.

27. I am indebted to Oliver Backhouse, VP of Strategic Partnerships at AstraZeneca, for providing the information in this paragraph.

28. https://www.nytimes.com/2021/03/16/world/europe/europe-astrazeneca-vaccine-suspensions.html, https://www.independent.co.uk/news/world/europe/covid-vaccine

-france-astrazeneca-macron-b1810072.html, and https://www.reuters.com/article/us-eu-summit-macron-idUSKBN2AP2TM, accessed August 5, 2021.

29. See ibid., p. 217 (as well as chapter 9 as a whole and p. 100). For example, in the Centered Leadership model that Keller and Schaninger present, "centeredness" includes mastering mental energy (thinking) and spiritual energy (being).

30. Among start-ups or financial companies, "unicorn" is a term for a privately owned business valued at more than $1 billion.

31. Whyte, 1994, pp. 96–97.

32. Https://www.thenational.ae/business/technology/hundreds-of-millions-spent-on-ai-as-will-i-am-tie-up-continues-majid-al-futtaim-chief-says-1.931098, accessed June 12, 2020. See also Keller and Schaninger 2019, pp. 151–152.

33. Https://www.thenational.ae/business/technology/hundreds-of-millions-spent-on-ai-as-will-i-am-tie-up-continues-majid-al-futtaim-chief-says-1.931098, accessed June 12, 2020.

34. Academic studies support the notion that purposeful thinking is more than just logical. In a 1995 *Journal of Personality and Social Psychology* paper, Professor Carol Ryff and Corey Lee Keyes of the University of Wisconsin-Madison posited a theoretical model of psychological well-being that includes six distinct, validated dimensions: Autonomy, Environmental Mastery, Personal Growth, Positive Relations, Purpose, and Self-Acceptance (Ryff and Keyes, 1995). Ryff and Keyes confirmed the fit and consistency of these dimensions using a wide-ranging sample of 1,108 adults (ibid., pp. 720–721). Several of the dimensions (especially Purpose) locate well-being beyond logical thinking.

35. https://bernardmarr.com/default.asp?contentID=1171; https://venturebeat.com/2017/07/11/coca-cola-reveals-ai-powered-vending-machine-app/.

36. For a discussion of the earlier transformation efforts, see Keller and Schaninger, 2019, p. 13, and https://www.irishtimes.com/business/retail-and-services/the-man-who-put-the-fizz-back-into-coca-cola-1.1617838, accessed June 24, 2020. See also https://www.intelligentautomation.network/transformation/articles/how-coca-cola-is-building-a-digital-first-business, accessed July 6, 2021.

37. From 2008, DTE Energy's teams intensified their focus on purpose, weaving it into onboarding and training programs, corporate meetings, and culture-building activities such as film festivals and sing-alongs. Professor Robert Quinn and Professor Anjan Thakor reported that DTE then "received a Gallup Great Workplace Award for five years in a row [and its] stock price more than tripled from the end of 2008 to the end of 2017" (https://hbr.org/2018/07/creating-a-purpose-driven-organization, accessed July 8, 2021). See also Malnight, Buche, and Dhanaraj (2019) and Malloch (2015).

38. Https://www.linkedin.com/pulse/accelerating-path-purpose-stakeholder-value-through-ai-karine-yuki, accessed July 8, 2021 (article cowritten by Estefania Molina, Tina Naser, and Karine Yuki).

39. Hagerty is project lead and researcher at the University of Cambridge Leverhulme Centre for the Future of Intelligence and the Centre for the Study of Existential Risk.

40. Lee, 2018, pp. 200.

41. On AI organizations and values, see https://openai.com/blog/ai-safety-needs-social-scientists/ (February 19, 2019, accessed July 16, 2020) and https://www.msn.com/en-gb/news/spotlight/the-case-for-taking-ai-seriously-as-a-threat-to-humanity/ar-AAB7D8h (May 9, 2019, accessed May 9, 2019).

42. Https://www.thoughtspot.com/data-chief/ep25/princeton-university-ruja-benjamin-on-bias-in-data-and-ai, accessed March 14, 2022.

43. "Optimism Is Associated with Exceptional Longevity in Two Epidemiologic Cohorts of Men and Women," Lewina O. Lee et al., *PNAS*, August 26, 2019, https://doi.org/10.1073/pnas.1900712116, accessed July 8, 2019.

CHAPTER 4

1. https://www.forbes.com/sites/daviddawkins/2019/08/29/where-are-the-aliens-jack-ma-and-elon-musks-weird-debate-on-space-ai-and-the-future-of-work/ and https://www.bbc.co.uk/news/technology-49508091, both published August 29, 2019, and accessed August 30, 2019.

2. According to research conducted by Apple in 2016, people unlock their iPhones on average eighty times per day (https://www.theverge.com/2016/4/18/11454976/apple-iphone-use-data-unlock-stats, accessed April 2, 2022). See also this 2018 research: https://www.statista.com/statistics/1050339/average-unlocks-per-day-us-smartphone-users/:, accessed April 2, 2022. This research indicates that those from Generation Z unlock their phones more than twice as often as do Baby Boomers.

3. Kline 2020, p. 92.

4. The fourteen forces are comprised of the ten Components of a Thinking Environment˚, plus the four systemic forces that Kline highlights in her 2020 book, *The Promise That Changes Everything: I Won't Interrupt You* (Penguin Random House).

5. Jacobs, 2017, p. 30.

6. *Am I Just My Brain?*, The Good Book Company, 2019.

7. Owen et al., 2006, p. 1402. The vegetative state is a disorder in which "patients who emerge from coma appear to be awake but show no signs of awareness."

8. Ibid.

9. Owen, 2017, Prologue. Owen made this observation after noting that 15 to 20 percent of people in the vegetative state are fully conscious. See also Monti et al.'s 2010 *New England Journal of Medicine* paper "Willful Modulation of Brain Activity and Communication in Disorders of Consciousness." (Professor Adrian Owen was one of the contributors.) In this paper, the authors found that four out of twenty-three patients classified as in the vegetative state were able to generate thoughts of tennis or their homes (pp. 3–4). One of these four patients gave the correct answers to a series of questions about his family, such as "Do you have any brothers?" Discussing this patient, Owen recalled that the team behind this article "were astonished when we saw the results of the patient's scan and that he was able to correctly answer the questions that were asked by simply changing his thoughts. . . . Not only did these scans tell us that the patient was not in a vegetative state but, more importantly, for the first time in five years it provided the patient with a way of communicating his thoughts to the outside world" (https://www.theguardian.com/science/2010/feb/03/vegetative-state-patient-communication, accessed October 21, 2010).

10. Materialism may also be defined as "a set of related theories which hold that all entities and processes are composed of—or are reducible to—matter, material forces or physical processes" (George J. Stack, in Craig [ed.], 2005, p. 633). Materialism is for many purposes the same as physicalism (see *Materialism in the Philosophy of Mind*, Howard Robinson, ibid., p. 635).

The kind of materialism that I describe in this section is perhaps what may be known as materialistic scientific naturalism.

11. On the historic pedigree of scientism, see Dirckx, 2019, p. 32.

12. Tallis, 2013, loc. 308.

13. Lennox, 2019, location 506.

14. Wilczek, 2019, p. 74. Wilczek views humans as currently having five "big advantages" over their artificial competitors: three-dimensionality; self-repair; integration of sensors and actuators; connectivity; and interactive development. Wilczek stated that the last two advantages are the most profound, in the sense of being furthest from AI's current capabilities. Among other things, Wilczek wrote the following about connectivity: "Human neurons typically support several hundred connections (synapses). Moreover, the complex pattern of these connections is very meaningful. Computer units typically make only a handful of connections, in regular, fixed patterns" (ibid., p. 72). On interactive development, Wilczek adds, "The human brain grows its units by cell divisions and orchestrates them into coherent structures by movement and folding. It also proliferates an abundance of connections among the cells. . . . Thus, the fine structure of the brain is tuned through interaction with the external world—a rich source of information and feedback!" (ibid., pp. 71–73). The two parts of this second quotation refer to two different but related things: how the brain organizes itself in a complex way, and how interaction with the external world influences this process. An example from child development may illustrate the relationship between these two things. My daughter is learning to play the violin. As she practices, her brain is growing and organizing. At the same time, her brain is being tuned through interaction with her teacher. At work here is neuroplasticity, a term given to the capacity of neurons to break old connections and reform new ones based on external stimuli. I am indebted to Dr. Sharon Dirckx, a former neuroscientist, for alerting me to the connection between the above two parts of Wilczek's observations about interactive development. Wilczek continues that connectivity and interactive development are synergistic (ibid., p. 74). That is to say, it is advantageous to make as many connections as possible. I am grateful to Dr. Holly Roy, Academic Clinical Fellow in Neurosurgery at the University of Plymouth, for pointing out to me that, while this last assertion is true, there is another important process that happens around adolescence called synaptic pruning. Synaptic pruning involves the number of connections being cut back and is thought to be a positive, rather than a negative, cognitive event, associated with brain maturation. See also https://www .scientificamerican.com/article/why-is-synaptic-pruning-important-for-the-developing -brain/, accessed October 21, 2020.

15. Ibid., p. 69. Wilczek's italics.

16. Okasha, 2016, p. 122.

17. As quoted by Hans Bohr in "My Father," published in *Niels Bohr: His Life and Work* (1967), p. 328. Source: https://quotepark.com/quotes/2057105-niels-bohr-the-opposite-of-every -great-truth-is-also-a-great/, accessed July 23, 2022.

18. Buolamwini and Gebru, 2018.

19. Https://www.thoughtspot.com/data-chief/ep25/princeton-university-ruja-benjamin-on-bias -in-data-and-ai, accessed March 14, 2022.

20. For example, in 2015, Google's photo service tagged black people as gorillas.

21. See a 2016 University of Washington study, reported here in 2017: https://www.wired.com /story/machines-taught-by-photos-learn-a-sexist-view-of-women/ (accessed July 12, 2021).

22. Https://www.technologyreview.com/2017/08/21/149585/ai-learns-sexism-just-by-studying
    -photographs/, accessed July 12, 2021.

23. Https://www.youtube.com/watch?v=rY8RkET3KC0, accessed March 18, 2022. Carina C.
    Zona observed that biases in photography- and AI-related data sets and processes have their
    roots in decisions made as early as the 1950s. At this time, Shirley Cards were introduced to
    help calibrate photographic developers, so that they could "accurately reproduce" details and
    colors (see https://www.youtube.com/watch?v=znwWYR1mzzw, accessed April 2, 2022).
    Shirley Cards increased accuracy especially in relation to white skin.

24. Https://www.technologyreview.com/2017/07/12/150510/biased-algorithms-are-everywhere
    -and-no-one-seems-to-care/, accessed July 12, 2021.

25. See discussion and references in chapter 3.

26. Appendix 4 explores an example of conflating machine purpose and intentional purpose,
    drawing on John Brockman's report of some of Stephen Wolfram's views.

27. Lennox, 2019, loc. 379. Lennox notes that "'[w]hy' questions connected with function, as
    distinct from 'why' questions associated with purpose, are usually regarded as within the
    domain of science" (ibid., loc. 1545).

28. Appendix 4 describes an example of an unhelpfully homogenized view of purpose, in an
    algorithmic context.

29. Kline 2020, pp. 117–118.

30. PublicAffairs, 2020.

31. Https://www.economist.com/open-future/2020/01/17/digital-disinformation-is-destroying
    -society-but-we-can-fight-back, accessed July 12, 2021.

32. Https://www.politico.eu/article/silicon-valley-is-losing-the-battle-against-election
    -misinformation/, accessed July 12, 2021.

33. Kline, 2020, p. 36.

34. Kline, 2015, p. 18.

35. "On the Far Side of Complexity," a video short filmed and produced by Lloyd Wigglesworth,
    https://www.timetothink.com/media/, accessed July 14, 2020.

CHAPTER 5

1. Https://positiveorgs.bus.umich.edu/wp-content/uploads/Leadership-When-Events-dont
   -Play-by-the-Rules-Weick.pdf, accessed August 8, 2020. See also William Schutz's aphorism
   "Understanding evolves through three phases: simplistic, complex, and profoundly simple"
   (*Profound Simplicity,* Los Angeles: Learning Concepts, 1979).

2. See chapter 2 for a discussion of Labyrinth. Also see the Labyrinth resource for practical tips.

3. The annotated numbered sections constitute a distilled subset of Thinking Environment*
   findings. For a fuller set of findings, refer to Kline's 2020 book, *The Promise That Changes
   Everything.* Close study of the thirty-seven management practices put forward by Keller and
   Schaninger shows that Thinking Environments* facilitate all nine elements and thirty-seven
   practices underpinning organizational health (Keller and Schaninger, 2019, pp. 39–42).

4. Wiseman, 2017, loc. 372.

5. Roberts and Wood, 2009, pp. 257–285, esp. pp. 261 and 277–279.

6. Brian Draper *Lent 40* email series of reflections, 2018, "The Way of Ma"; also https://wawaza .com/blogs/when-less-is-more-japanese-concept-of-ma-minimalism-and-beyond/, accessed July 17, 2020.

7. Kline 2020, pp. 170–172.

8. The interrupting cow joke has been a staple of dad jokes for years. It goes like this: *Knock knock. Who's there? Interrupting cow. Interrupting c . . . MOOOO-OOOOOO.* My daughters love this joke and had fun diversifying it to many other animals.

9. Hillary Rodham Clinton, *What Happened*, London: Simon & Schuster UK Ltd., 2017, p. 137.

10. Bienkowski 2010, p. 40–41.

11. Scharmer, 2016, loc. 972. Scharmer's four social fields are relevant here: I-in-me: acting from the center inside one's organizational boundaries (field 1: habitual). I-in-it: acting from the periphery of one's organizational boundaries (field 2: subject-object). I-in-you: acting from beyond one's organizational boundaries (field 3: relational). I-in-now: acting from the emerging sphere across one's open boundaries (field 4: generative) (ibid., Loc. 4,138).

12. McGilchrist, 2021, loc. 435. McGilchrist defined attention as "the manner in which our consciousness is disposed towards whatever else exists" (ibid., loc. 440).

13. Daniel Goleman puts it as follows: "A primary task of leadership is to direct attention. To do so, leaders must learn to focus their own attention" ("The Focused Leader," *Harvard Business Review*, December 2013, pp. 52). For Goleman, focusing attention includes commanding "the full range of their own attention," which is "a triad: an inward focus, a focus on others, and an outward focus" (ibid., pp. 53 and 60).

14. Kline, 2015, p. 76.

15. Jacobs, pp. 44, 84, and 87.

16. Kline, 2015, p. 53.

17. "Accelerating Analytics to Navigate COVID-19 and the Next Normal," May 21, 2020, https://www.mckinsey.com/business-functions/mckinsey-analytics/our-insights/accelerating -analytics-to-navigate-covid-19-and-the-next-normal, accessed July 23, 2020. See also Keller and Schaninger, 2019, pp. 41 and 85.

18. Prof. Paul Brown, lecture to the *Time to Think* Collegiate, October 2011). For other relevant research, visit https://www.timetothink.com/thinking-environment/research-case-studies/.

19. Kline, 2020, p. 56.

20. See also Martin's discussion of ways that integrative thinkers such as Lafley, Young, and Handling face unpleasant trade-offs (2009, loc. 953).

21. Alan Lafley and Roger Martin, *Playing to Win: How Strategy Really Works*, Boston: HBR Press, 2013, p. 136.

22. Radecki and Hull, 2018, p. 63.

23. Ibid., p. 82.

24. See, for example: "The Role of Positivity and Connectivity in the Performance of Business Teams: A Nonlinear Dynamics Model," Marcial Losada and Heaphy, Emily, *American Behavioral Scientist*, Vol. 47 (6), 740–765, 2004; and https://www.gottman.com/blog/the-magic -relationship-ratio-according-science/, accessed July 16, 2020.

25. "Doing the Right Thing Is Proving Harder Than Google Staff Had Hoped," Simon Duke, *The Times* (of London), April 25, 2019.

26. https://openai.com/blog/ai-safety-needs-social-scientists/, February 19, 2019, accessed July 16, 2020.

27. Https://www.msn.com/en-gb/news/spotlight/the-case-for-taking-ai-seriously-as-a-threat -to-humanity/ar-AAB7D8h, Kelsey Piper, May 9, 2019, accessed May 9, 2019.

28. Https://ainowinstitute.org/research.html, accessed May 9, 2019.

CHAPTER 6

1. I am indebted to Surya Ramkumar for suggesting this example.

2. http://www.jamesmumford.co.uk/package-deal-ethics-2/, accessed January 28 2019.

3. "Fei-Fei Li's Mission to Transform Healthcare AI," *HBR Exponential View* podcast (with Azeem Azhar), May 13, 2020.

4. Kline, 2020, p. 36.

5. Kounios, 2014 and Radecki et al., 2018, p. 56.

6. I am grateful to Prof. Peter Thonemann, Professor of Ancient History at the University of Oxford, for confirming that this quote is not from Ovid (b. 43 BC), contrary to frequent claims on the internet and elsewhere. The earliest reference to this quotation we can find is to Charles Brower, in the 1964 edition of the Michigan Technic, vols. 81–83, p. 40.

7. On food waste, see https://www.mckinsey.com/business-functions/sustainability/our -insights/sustainability-blog/how-ai-can-unlock-a-127b-opportunity-by-reducing-food -waste. On climate, see https://www.nationalgeographic.co.uk/science-and-technology /2019/07/how-artificial-intelligence-can-tackle-climate-change and https://blogs.ei.columbia .edu/2018/06/05/artificial-intelligence-climate-environment/ (all accessed August 14, 2020).

8. "Why Governments Need an AI Strategy: A Conversation with the WEF's Head of AI," McKinsey publishing, In conversation with David DeLallo, May 26, 2020, https://www.mckinsey .com/business-functions/mckinsey-analytics/our-insights/why-governments-need-an-ai -strategy, accessed March 20, 2022.

9. Https://www.thevintagenews.com/2018/06/08/kitty-genovese/, accessed July 14, 2021.

10. See chapter 3 for a description of my experience leading Coachify, a wearables, coaching, and AI company.

11. See Radecki et al. 2018, https://www.nytimes.com/2016/02/28/magazine/what-google -learned-from-its-quest-to-build-the-perfect-team.html/ and https://hbr.org/2017/08/high -performing-teams-need-psychological-safety-heres-how-to-create-it, accessed July 15, 2020.

12. 2015, p. 76 and personal conversation.

13. A "round" is a meeting process in which each person gets one and only one chance to speak, in sequence. No one speaks until everyone has had a chance to speak, unless it's your turn to speak.

14. Roberts and Wood, 2009, pp. 257–285, esp. pp. 261 and 277–279. See also Keller and Schaninger's treatment of the link between social relationships and leadership effectiveness (2019, pp. 222f).

15. Keller and Schaninger, 2019, pp. 161–166 and 222f.

16. This research is described in Kline 2015, pp. 291–294. Havers's research related to the Transforming Meetings Program and involved interviews with fifteen senior officers from

eleven organizations on three continents. Whitehead's research related to the *Time to Think Mentoring Program*, involving sixteen mentoring pairs.

17. Waldock, 2011, back cover.

18. Willis Towers Watson describe the following six themes that impact the evolution of corporate governance models: gig-economy governance, tech ethics, cyber security and data protection, skills and organizational development, organizational culture and human capital management, and from fiduciary to advisory (https://www.willistowerswatson.com/en-GB /Insights/2019/08/the-future-of-work-and-its-impact-on-corporate-governance, accessed July 16, 2020). See also the report, "A Government Blueprint to Adapt the Ecosystem to the Future of Work" (https://www.mckinsey.com/industries/public-sector/our-insights/a-government -blueprint-to-adapt-the-ecosystem-to-the-future-of-work, accessed July 16, 2020).

19. According to Professors Nicholas Epley and Amit Kumar, progress in AI organizational ethics requires moving beyond legalistic approaches that merely punish "bad apples," to creating ethical cultures that address values, incentives, norms, and what are called "thoughts during judgment" (2019, pp. 144ff).

20. Waldock, 2011, pp. 46–48.

21. Boris Babic et al., 2020, p. 58.

22. Design Thinking is a five-step, problem-solving process. Roger Martin's knowledge funnel involves exploring before exploiting. According to Martin, "The velocity of movement through the knowledge funnel, powered by design thinking, is the most powerful formula for competitive advantage in the twenty-first century." Discovery Driven Planning is a process for taking an ongoing learning approach to strategy. In a digital context, this shifts the question subtly from "What new business model should we back?" to the more incremental, "How can we learn our way toward a model that's right for our business?" (McGrath and McManus, 2020, p. 127). The approach of Keller and Schaninger to innovation and learning draws on their research about leading large-scale change (Keller and Schaninger, 2019, p. 42).

23. McGrath and McManus, 2020, p. 130.

24. Ibid.

CHAPTER 7

1. These words are a slight adaptation of MacFarquhar's paraphrase of something that the roboticist Rodney Brooks said (MacFarquhar, 2018).

2. In a pleasing twist, that lessens my embarrassment about this story not one iota, I discovered something in the attic a few months after writing this story: my Presidential term card, where I found the required information. I can now somewhat sheepishly report that the individual with whom I dined was James Watson.

3. For the purposes of this definition, in using the word "brain" I am not taking a view on the relationship between the mind and the brain. See discussion in chapter 4.

4. 2021, loc. 376.

5. In her perceptive *New York Times* article about the "mind-expanding ideas" of philosopher and cognitive scientist Andy Clark, journalist Larissa MacFarquhar explored the idea of cognition as embodied language, as well as related ideas relevant to embodied intelligence (MacFarquhar, 2018).

6. Https://strozziinstitute.com/embodied-leadership-trainings/, accessed July 22, 2021.

7. In the context of embodiment and therapy, I also recommend books by Bessel van der Kolk (2014), Ogden and Fisher (2015), McBride (2021), and Taylor (2021).

8. "The Soul's Tomb: Plato on the Body as the Source of Psychic Disorders," published online November 16, 2020, Dr. Douglas R. Campbell, https://www.degruyter.com/document /doi/10.1515/apeiron-2020-0022/html, accessed July 19, 2021. Platonism also influenced a collection of religious ideas known as Gnosticism, which is also relevant here as Gnosticism generally holds that truth exists in the mind, rather than the body, which is subject to decay and death.

9. "Centering Prayer," *Nomad Podcast*, Ep. 199, released June 20, 2019.

10. Https://hbr.org/2001/01/the-making-of-a-corporate-athlete, accessed July 16, 2021.

11. See for example *The Corporate Athlete: How to Achieve Maximal Performance in Business and Life*, Dr. Jack Groppel (Wiley, 1999).

12. Schwartz, Gomes, and McCarthy, 2016 reissued edition. See also Schwartz and Gomes, 2010.

13. On human rhythms, see Oakley, 2017. On sleep, see Walker, 2018.

14. Schwartz, Gomes, and McCarthy position physical, mental, and emotional energy as the "what" and spiritual energy as the "why" (2010, p. 238). To some extent, all four pathways of threshold leadership address both the "what" and the "why."

15. Gardner, 2006, pp. 9-10.

16. *The Hungry Spirit: Beyond Capitalism—A Quest for Purpose in the Modern World* (Broadway Books, 1999).

17. Schwartz, Gomes, and McCarthy, 2010, pp. 19–20.

18. Bethany McLean and Peter Elkind, *The Smartest Guys in the Room: The Amazing Rise and Scandalous Fall of Enron*, New York: Penguin, 2013.

19. "Are Trees Human?" *BBC* podcast, December 2018.

20. Https://www.psytoolkit.org/experiment-library/stroop.html, accessed October 21, 2020.

21. Schwartz, Gomes, and McCarthy, 2010.

22. Tugade, M. M., & Fredrickson, B. L. (2004). "Resilient Individuals Use Positive Emotions to Bounce Back from Negative Emotional Experiences," *Journal of Personality and Social Psychology*, 86(2), 320–333 (https://doi.org/10.1037/0022-3514.86.2.320).

23. McBride, 2021, loc. 140.

24. Https://positivepsychology.com/positive-emotions-list-examples-definition-psychology/, accessed July 3, 2020. Also see this peer-reviewed study, in which twenty-seven distinct emotions exist: "Varieties of reported emotional experience," Alan S. Cowen, Dacher Keltner, *PNAS*, September 2017, 201702247; DOI: 10.1073/pnas.1702247114.

25. Lewina O. Lee et al., "Optimism Is Associated With Exceptional Longevity in 2 Epidemiologic Cohorts of Men and Women," *PNAS*, August 26, 2019, https://doi.org/10.1073 /pnas.1900712116, accessed July 26, 2021.

CHAPTER 8

1. See references in Schwartz, Gomes, and McCarthy, 2010. In 2019, the *Harvard Business Review IdeaWatch* column reported an interesting contrarian view. In summarizing latest

research on multitasking, *IdeaWatch* noted the following: "Even though cognitive scientists know there is no such thing, although fascinatingly, latest research shows that if you think you are multitasking, you will do better" (Vol. 97, Issue 5, September–October 2019, p. 24).

2. Zuboff, 2019, pp. 464f.

3. Many other social media companies, including Twitter and TikTok, also use machine learning in their core processes, including facial recognition in photos, targeting users with advertising, and selecting attractive content to display.

4. Physical energy comprises four components: nutrition, exercise, daytime rest, and sleep (Schwartz, Gomes, and McCarthy, 2010, chapters 5–8).

5. Https://edition.cnn.com/2012/10/14/us/skydiver-record-attempt/index.html, accessed July 16, 2021.

6. Https://www.webcitation.org/6F7ZfaO2M?url=http://www.fai.org/fai-slider-news/37012 -baumgartners-records-ratified-by-fai, accessed July 16, 2021.

7. Https://www.today.com/news/felix-baumgartner-i-didnt-enjoy-space-jump-flna1C6601083, accessed July 16, 2021.

8. Walker, 2018, p. 133.

9. According to Matthew Walker, peer-reviewed studies link sleep loss to the following conditions: Alzheimer's disease, anxiety, depression, bipolar disorder, suicide, stroke, chronic pain, cancer, diabetes, heart attacks, infertility, weight gain, obesity, and immune deficiency (Walker 2018, p. 133). This list is illustrative, not exhaustive, although reading it is exhausting.

10. Walker, 2018, p. 133.

11. "Poetry and the Journey of the Soul," *Nomad Podcast*, N225.

12. Feldman Barrett, 2017, p. 293.

13. Ibid., Introduction (loc. 90).

14. Reported by Howard Gardner (2006, p. 9).

15. Gardner, 2006, p. 8.

16. "China's AI Boom," Kai-Fu Lee, *Time*, August 23 / August 30, 2021, p. 68.

17. For example, an artificially intelligent machine may incorporate the outputs of operations on earlier versions of a training set, into how it operates in later iterations.

18. In Brockman (ed.), 2019, p. 230.

19. Gardner, 2006, p. 8.

20. In Brockman (ed.), 2019, p. 230.

21. McBride 2021, locs. 202 and 3,530. The benefits of body movement, described in the latter reference, arise from the fact that muscular contraction releases short chains of amino acids called myokines, which then travel to the brain.

22. In 1982, van der Kolk also founded the Trauma Center (now the Trauma Research Foundation), while a junior faculty member at the Harvard Medical School.

23. Van der Kolk drew on disciplines including neuroscience, developmental psychopathology ("the study of the impact of adverse experiences on the development of mind and brain"), and interpersonal neurobiology ("the study of how our behavior influences the emotions, biology, and mind-sets of those around us"). See van der Kolk, 2014, pp. 2 and 10.

24. Ibid., p. 2.

25. AlphaGo Zero superseded AlphaGo, the program that in 2016 defeated the eighteen-time Go world champion Lee Sedol by a score of 4-1.

26. Https://venturebeat.com/2019/11/20/deepminds-muzero-teaches-itself-how-to-win-at-atari-chess-shogi-and-go/, accessed July 20, 2021.

27. Https://www.santafe.edu/research/projects/theory-of-embodied-intelligence, accessed May 20, 2021.

28. A form of this reflection first appeared in 2018, as part of Brian Draper's Advent reflection series, "God breathed."

29. See the trailer here: https://www.youtube.com/watch?v=JycCFypvgmI, accessed March 9, 2022.

30. https://www.healthywa.wa.gov.au/Articles/A_E/Calming-techniques-breathing-training#:~:text=Anxious%20breathing,and%20how%20deeply%20we%20breathe, accessed August 5, 2021.

31. Walker, 2018, locs. 4,283–4,297.

32. Ibid., loc. 506.

33. *Lent 40* email series of reflections, 2021, "Peace Is Every Step."

34. https://news.utexas.edu/2017/06/26/the-mere-presence-of-your-smartphone-reduces-brain-power/, accessed July 16, 2021.

35. Paula Mariani, "Embodied Leadership," October 18, 2019, in *To Be Radiant*, https://medium.com/to-be-radiant/embodied-leadership-681bd89d2eaf, accessed July 22, 2021.See also Pete Hamill, *Embodied Leadership: The Somatic Approach to Developing Your Leadership* (Kogan, 2013).

36. Van der Kolk, 2014, loc. 4747.

CHAPTER 9

1. "High performing" refers to organizations that best align around and achieve strategic goals. See Claudio Feser, Michael Rennie, and Nicolai Chen Nielsen, *Leadership at Scale* (Boston: Hachette, 2018) and Keller and Schaninger, 2019.

2. Iflexion, "Why Do AI Projects Fail Too Often?" January 19, 2021, *Becoming Human: Artificial Intelligence* magazine, https://becominghuman.ai/why-do-ai-projects-fail-too-often-4c4f973bb195, accessed July 26, 2021.

3. I am indebted to Georgie Lyttleton for alerting me to these words of Herbert Spencer, which, in various forms, have also been attributed to Paley and Einstein. See also Jacobs (2017, p. 146). Jonathan Haidt's notion of six "moral tastebuds" or six "taste receptors of the righteous mind" also repays study in this ethical context; see *The Righteous Mind: Why Good People Are Divided by Politics and Religion* (New York: Vintage Books, 2013).

4. Henderson-Espinoza, 2022, pp. 9 and 18.

5. In an interesting coda that fascinates geeks like me, a debate has been raging in the last twenty-four months about whether humans realistically could supply enough power for robots. For opposing views, see https://www.esquire.com/entertainment/movies/a26978756/the-matrix-human-batteries-plot-hole-explained/ and https://www.syfy.com/syfywire/science-behind-the-fiction-humans-as-batteries-as-in-the-matrix-probably-not-gonna-happen, accessed July 20, 2021.

6. Jemima McAlpine, while interviewing Damon Garcia, "The God Who Liberates," *Nomad Podcast*, N247.

7. Baker 2019, p. 1. Used with permission.

8. Https://link.springer.com/article/10.1007/s40926-018-0087-0, accessed July 26, 2021.

9. Scharmer 2016, locs. 191–199. Scharmer applies his theory to four systemic levels of micro (individuals), meso (groups), macro (institutions), and mundo (ecosystems) (ibid., loc. 312).

10. For example, Scharmer connects thinking and being as follows: his fourth and most profound level of listening (which is part of thinking) is "My whole being slowed down . . . I am connected to something larger than myself" (Scharmer, 2016, loc. 989). Scharmer also proposes four primary actions "through which human beings collectively enact the social field: attending, conversing, organizing, and coordinating," actions via which participants optimally connect their thinking and their being (ibid., loc. 4217).

11. Scharmer 2016, loc. 3490.

12. Ibid., locs. 3571–3664. The fourth principle, which precedes the other three, is "The power of intention." Pathways 1 and 2 address this. For the embodied examples, see ibid., locs. 1701–1703, 2993, and 3385.

13. Keller and Schaninger, 2019, pp. 146–148.

14. For fuller descriptions of the nine intelligences at the end of this list, see Gardner, 2006, pp. 8–21.

CHAPTER 10

1. Ramakrishnan defines AGI as "a general-intelligence machine that will think like a human and possibly develop consciousness" (in Brockman [ed.], 2019, p190). This is closer to Bostrom's definition of Superintelligence and to Tegmark's definition of "General Intelligence" (the "[a]bility to accomplish virtually any goal, including learning," 2018, p. 30), than to Tegmark's definition of AGI. See also Lee, 2018, pp. 140–142.

2. A more granular vocabulary of advanced AI could further distinguish levels of machine competency within each Age of AI. When discussing Superintelligence that operates far beyond human levels, even in emotions and wisdom, I prefer to use the term Advanced Total Intelligence (ATI). For Superintelligence that operates far beyond human levels, Stuart Russell uses the terms superhuman AI or Artificial Super Intelligence (ASI; "substantially more capable than humans across all relevant dimensions," "Beneficial Artificial Intelligence," an *HBR Exponential View* podcast, June 26, 2019). I've also seen references to "Artificial Total Intelligence (AGI)," in a review of Bostrom's book *Superintelligence* and in articles on Bitcoin and Google's RankBrain—see for example https://content.chineseculturehub.com /en/ittc/6844660195531424772.html?人工智能对SEO的改变超乎你的想象!, accessed March 28. 2019.

3. Pinker, 2019, p. 110.

4. Pearl, 2019, p. 16.

5. Russell, 2019, pp. 7–8.

6. Seth Lloyd pursues a similar theme (2019, p. 8).

7. Tegmark, 2018, p. 31.

8. Pinker, 2019, pp. 109–110.

9. Russell, 2019, p. 73, https://engineering.berkeley.edu/brett/, accessed July 28, 2020.

10. Https://engineering.berkeley.edu/brett/, accessed July 28, 2020. One other noteworthy robot dexterity project is OpenAI's Dactyl robot hand (https://openai.com/blog/learning-dexterity/, accessed July 28, 2020).

11. Bienkowski, 2010, p. 68.

12. Heifetz's full academic title is the King Hussein bin Talal Senior Lecturer in Public Leadership at the John F. Kennedy School of Government, Harvard University.

13. See Kegan and Lahey, 2009, pp. 29–31.

14. For example, see Kegan, 1982 and 1994, and Kegan and Lahey, 2001, 2009, and 2016. In the context of leadership, Kegan and Lahey point out that in using the term *development*, they mean "not the development of a career, but the development of the person having the career" (2016, p. 58). Other relevant developmental models include those of Frederick Laloux (organizational development), Clare Graves (Spiral Dynamics), Mary May ("Whirl" model of consciousness development), Jean Piaget (cognitive development), Ken Wilber (integral approach), Jean Gebser (structures of consciousness), Lawrence Kohlberg (model of moral development), William Perry (model of intellectual and ethical development), Jane Loevinger (model of ego-state development), Bill Torbert (action-logics), and Susanne Cook-Greuter (Ego Development Theory, EDT). These models have much to offer and overlap with at least parts of ADT. For example, Wilber draws together multiple ways of looking at the world, together with developmental levels, multiple intelligences, states of consciousness, and more (2011, introductory note and pp. 2, 8–20, 33, and 207–208). In addition, Kegan is not without his critics. For example, psychologists Michael Basseches and Michael Mascolo claim that Kegan's and others' stage development models underemphasize social and cultural relations (Basseches and Mascolo, 2009, p. 32). However, I do not think that such emphasis was Kegan's aim. Another useful model relating to complexity, which, unlike those listed earlier in this footnote is not a stage development model, is the Cynefin framework. The Cynefin framework is a dynamic sense-making model created by Dave Snowden. "Cynefin" is a Welsh word that signifies the multiple factors in our environment that influence us in ways we don't understand. The framework allows shifts in the following five decision-making domains: Simple (known knowns; causality is repeatable, predictable, and self-evident); Complicated (known unknowns in which causality is repeatable and predictable, but understandable only with analysis); Chaotic (Unknowable unknowns without discernible causality); Complex (Unknown unknowns, where causality exists, but is only obvious in hindsight), and Disorder (the domain of not knowing which of the first four domains applies). The Cynefin framework proposes corresponding approaches and practices to decision-making in each of these domains, allowing leaders to explore their perceptions, analyze behavior, and make decisions. For more information, see "The New Dynamics of Strategy: Sense-Making in a Complex and Complicated World," D. Snowden, *IBM Systems Journal*, 2003.

15. Kegan and Lahey, 2009, pp. 11–13.

16. Kegan, 1984 and 1992, and Kegan and Lahey, 2009.

17. Kegan provides the following longer description of the self-transforming stage: "There can come a time in your own evolution where you come to recognize that as powerful as your own inner compass enables you to be, inevitably your system for making meaning with which you've become identified, has limitations, has blind spots, leaves something, privileges something and disadvantages something else. As you come to see those limits, you have the potential now to again step away from something you've been completely embedded in. That is the gradual move of development, that you move from a subject to a way of making

meaning, to being able to step back from it and turn it into object, and look at it, and be in some relationship to it, and you begin to construct a new subjectivity, one which in the Wilberian phrase, transcends but includes the prior way of knowing" (Robert Kegan, "The Evolution of the Self," an interview with David Fuller, The Rebel Wisdom YouTube channel, May 31, 2019, https://www.youtube.com/watch?v=bhRNMj6UNYY, accessed August 3, 2021).

18. This fourth pathway of maturing consciousness rests on the two powerful ideas of constructivism and developmentalism. "Constructive" means creating conditions for each person to make meaning of things and to create a more thriving world, rather than merely accepting a world outside that we discover or have imposed on us. "Developmental" means creating environments for your team to grow in the sense of becoming more and more complex, not just in the sense of getting bigger. See Garvey Berger, 2012, p. 15 and Robert Kegan, "The Evolution of the Self," an interview with David Fuller, The Rebel Wisdom YouTube channel, May 31, 2019, https://www.youtube.com/watch?v=bhRNMj6UNYY, accessed August 3, 2021.

19. Jennifer Garvey Berger wrote something similar: "Developmental theories help us examine the match between a person's form of mind at any given time and the form of mind that may be required by a particular task " (2012, p. 14).

20. Kegan and Lahey, 2016, p. 60.

21. Kegan and Lahey hint at this connection between thinking and being. For example, they advocate shifting from habitual and unreflective patterns to more deliberate and self-reflective patterns by asking such questions as, "What anxieties and big assumptions does our doing or not doing imply?" (2009, pp. 48–49, 77–79, 202, 215, and 246–251). This constitutes thinking about doing in a way that connects with other aspects of our being, such as fears and beliefs. Throughout their work, Kegan and Lahey promote thinking about personal and organizational change in a way that is deeply rooted in a sense of self and thereby connects thinking and being at individual and organizational levels.

22. Kegan, 1994, p. 5.

23. Brian Draper, *Lent 40* email series of reflections, 2019.

CHAPTER 11

1. The following examples from Josephs and Joiner map the first three traits to threshold-equivalent leader levels: humility—"genuine dedication to the common good" (Josephs and Joiner's "cocreator" category); ease with tension—"empathetic awareness of conflicting stakeholder interests" ("synergist"); play—"playfully familiar" ("synergist"). The word *beneficial* echoes Stuart Russell's three principles of Beneficial AI, in which machines are "beneficial to the extent that their actions can be expected to achieve our objectives." The three principles are as follows: 1. Purely altruistic machines: The machine's only objective is to maximize the realization of human preferences; 2. Humble machines: The machine is initially uncertain about what those preferences are; 3. Learning to predict human preferences: The ultimate source of information about human preferences is human behavior (Russell, 2019, pp 9, 11, 159 and 171–179). Hannah Fry, Associate Professor in the Mathematics of Cities at the Centre for Advanced Spatial Analysis at University College London, also seems to have advocated some similar principles to Russell (Fry, 2018, loc. 3,051). It remains to be seen whether the required mathematics can be worked out to sustain Russell's or Fry's ideas and

whether a critical mass of beneficially minded commercial, political, and other actors grasp the opportunity to work this out before less beneficially minded ones do.

2. Roberts and Wood, 2009, p. 241. On humility as a developmentally mature quality, see Kegan and Lahey, 2016, pp. 172 and 176.

3. Roberts and Wood, 2009, p. 254. The quotation is from *Galileo in Rome*, Shea and Artigas, New York: OUP, 2003, p. 26 (see also p. 78).

4. Roberts and Wood, 2009, p. 254.

5. Swedish philosopher and economist Professor Erik Angner defines epistemic humility as "the realization that our knowledge is always provisional and incomplete—and that it might require revision in light of new evidence." "Epistemic humility—knowing your limits in a pandemic," Behavioral Scientist, April 13, 2020, behavioralscientist.org. Leading business thinkers recognize this provisionality to be a vital problem-solving mindset (Conn and McClean, 2020). See also Roberts and Wood's *Intellectual Virtues*, especially chapter 6, "Love of Knowledge" (2009). This latter work functions as an excellent philosophical introduction to the topic of humility. On provisionality of knowledge: I can envisage AGI showing provisionality of knowledge. One of Stuart Russell's principles of Beneficial AI is "humble machines." For Russell, this principle involves creating machines that defer to humans and tend to avoid single-mindedness. Such machines would exhibit uncertainty about human preferences in such a way that the machine would allow itself to be switched off (Russell, 2019, pp. 175–176.).

6. See "The Biggest AI Risks: Superintelligence and the Elite Silos," Ben Goertzel, https://www.youtube.com/watch?v=LvdSIdILCAo, accessed August 7, 2020.

7. Https://www.bruegel.org/2021/10/concentration-of-artificial-intelligence-and-other-frontier-it-skills/, accessed March 17, 2022. See also Lee 2018, pp. 21 and 169.

8. Goertzel, 2015. Goertzel was quoting others' views, not stating his own. See also Bostrom's Vulnerable World hypothesis, in which he proposes increased government-controlled surveillance as a counter to what he, ill-advisedly, calls "black-ball" hypotheses where individuals kill hundreds of millions of people, https://www.nickbostrom.com/papers/vulnerable.pdf, accessed August 7, 2020.

9. Https://www.businessgrowthhub.com/manufacturing/resources/blog/2020/05/the-power-of-collaboration-post-covid, https://www.wionews.com/opinions-blogs/collaborate-not-compete-lessons-from-the-new-normal-330324, https://www.startupdaily.net/2020/04/how-collaborating-with-other-businesses-helps-everyone-through-coronavirus/, and https://www.bethebusiness.com/productivity-insights/coronavirus-how-to-collaborate-with-other-businesses/, accessed October 25, 2020.

10. "The Age of Continuous Connection", *HBR*, May–June 2019.

11. To derive this number, I assumed the global human population to be eight billion, based on the current actual population of 7,933,545,673 and allowing for some growth between now and this book's release date, https://www.worldometers.info/world-population/, accessed at 08:34 GMT on Tuesday, March 15, 2022.

12. Fry, 2018, loc. 1185 and Kasparov, 2017, p. 234.

13. Kegan, 1982, p. 239.

14. "Beneficial Artificial Intelligence," an *HBR Exponential View* podcast, June 26, 2019.

15. See also Pinker in Brockman (ed.), 2019, p. 107.

16. Kahneman, 2011, chapter 5 (especially p. 70).

17. Kegan and Lahey, 2016, pp. 85 and 168.

18. See https://jrodthoughts.medium.com/learning-by-competition-understanding-adversarial -neural-networks-part-i-b8dab1b99b75, accessed May 29, 2019.

19. The two types of networks were discriminator networks and generator neural networks.

20. See chapter 4 and Appendix 4 for a discussion and example of mechanistic and intentional purpose. "Functional" is the equivalent to "mechanistic" in this chapter.

21. Genetic algorithms, another recent and promising avenue of AI research, suffers from the same problem. As part of the five phases of creating a genetic algorithm, someone needs to give the algorithm the inputs (presumably with a desired goal in mind); otherwise the exercise won't start. The five phases of creating a genetic algorithm are: 1. Creating an Initial population; 2. Defining a Fitness function; 3. Selecting the parents; 4. Making a Crossover; 5. Mutation (Manish Kumar, https://medium.com/analytics-vidhya/understanding-genetic -algorithms-in-the-artificial-intelligence-spectrum-7021b7cc25e7, accessed May 29, 2019). One of Russell's principles for Beneficial AI is nurturing AIs that learn to predict human behavior and core to this principle is the idea of potentially changing human preferences (Russell, 2019, pp. 240–245).

22. Bienkowski, 2010, pp. 16–18, 24–25, and 32.

23. Bienkowski, 2010, p. 53.

24. "Understanding the Leader's 'Identity Mindtrap': Personal Growth for the C-Suite," *McKinsey Quarterly*, January 2020.

25. Ibid.

26. Jacobs, 2017, p. 58.

27. D. Neale, K. Clackson, S. Georgieva, et al. (2018). "Towards a Neuroscientific Understanding of Play: A Dimensional Coding Framework for Analyzing Infant-Adult Play Patterns," *Frontiers in Psychology, 9*, p. 273.

28. See Dave Neale's discussion of the developmental benefits of play (https://thepsychologist. bps.org.uk/volume-33/may-2020/golden-age-play-adults, accessed August 7, 2020). Many of Csikszentmihaly's descriptions of the ideal flow state pertain to playful activities such as skiing, fishing, or playing the guitar (1992).

29. Https://mres.medium.com/the-playfulness-of-anne-frank-e49dc18ce3fd, accessed March 16, 2022.

30. Wilson and Daugherty, 2018.

31. Quoted in Brown, 2010, p. 101.

32. Scharmer quotes W. Brian Arthur, founder of the Economics Program at the Santa Fe Institute, as saying that some leading Chinese and Japanese artists will "sit on a ledge with lanterns for a whole week, just looking, and then suddenly they'll say, 'Ooohh' and paint something very quickly" (2016, loc. 1254). This fusion of contemplation and action strikes me as a mature kind of play.

33. Berger and Achi, "Creative Futures," a pre-publication chapter shared privately. See also William Turner, "even in the ludic play domain of certain of our liminal moments: play is more serious than we, the inheritors of Western Puritanism, have thought" (1969, pp. vii–viii).

34. Https://neurosciencenews.com/stress-playful-people-5918/amp/, accessed September 27, 2020.

35. Van der Kolk 2014, loc. 3,835.

36. See https://www.mckinsey.com/business-functions/organization/our-insights/the-psychology
    -of-change-management, accessed July 13, 2020.

37. Berinato, 2020. On the Six Stages of Grief model, see Kessler, 2019. The first five stages of
    grief are relatively well known, being based on the work of Elizabeth Kübler-Ross. Kessler
    worked with Kübler-Ross and built on her findings, adding a sixth stage, "finding meaning."
    This last stage does not involve thinking "this thing that happened to me was not, in fact,
    bad, but good." Rather it involves "finding meaning in the darkest hours (now)" and "finding
    meaning in the life that you will gradually build (in future)."

38. I am grateful to Tor Mesøy and Jennifer Garvey Berger, and to the faculty at the Conversa-
    tions at the Growing Edge program. Their inputs generated many of the questions in this
    section.

39. Berger and Achi, "Creative Futures," a pre-publication chapter shared privately and with
    permission.

40. Vernice Jones inspired some of these questions, during an adult development program she
    co-led.

41. See Babic et al., 2020, p. 58.

CHAPTER 12

1. Kegan, 1984, p. 130.

2. Dirckx, 2019, p. 40.

3. Chapman, 1992.

4. Https://hbr.org/2015/02/what-ceos-are-afraid-of, Roger Jones, accessed November 23, 2020.

5. Sanders, 2012, p. 61.

6. See, for example, "in the Seventh Story, human beings are not the protagonist of the
   world . . . Love is" (McLaren and Higgins, 2019, p. 9).

7. Ibid., p. 10.

8. Ibid., p. 25.

9. Https://www.scientificamerican.com/article/why-we-are-wired-to-connect/, accessed August
   21, 2021.

10. Https://www.goodreads.com/quotes/183896-man-is-by-nature-a-social-animal-an
    -individual-who, accessed September 27, 2020.

11. Https://www.huffingtonpost.co.uk/entry/japan-loneliness-aging-robots-technology
    _n_5b72873ae4b0530743cd04aa and https://www.telegraph.co.uk/technology/2019/01/08
    /tech-companies-target-ageing-baby-boomers-robot-pets-gadgets/, accessed September 27,
    2020.

12. Https://www.independent.co.uk/life-style/gadgets-and-tech/features/robot-carer-elderly
    -people-loneliness-ageing-population-care-homes-a8659801.html, accessed September 27,
    2020.

13. Https://www.youtube.com/watch?v=KAuTq86uShA, accessed September 27, 2020.

14. See Russell's survey of Nobel Prize–winning economist Daniel Kahneman's discussion of
    experiments about the experiencing self and the remembering self. One experiment con-
    cerns a subject's experience of having their hand plunged into cold water (Russell, 2019,
    pp. 238–240). See also chapter 1 for a discussion of consciousness and qualia.

15. Waldock, 2011.

16. Https://www.bbc.co.uk/bbcthree/article/c62bcab6-db6f-4026-90bb-7f508705a65b, accessed March 16, 2022. I am not a fan of the terms *neurodiverse* and *neurodivergent*, as they imply that some humans have diversity or divergence and others do not. In fact, everyone's brain is different to (diverse in relation to; divergent from) everyone else's brain, so we are all equal in this regard, which the terms do not imply. Those experts in diversity that I have consulted about this topic agree with these points and, like me, can't think of a better term. Equally, I appreciate that some labels can be helpful, especially for groups who have historically been marginalized or under-served.

17. Https://www.bbc.co.uk/news/uk-england-birmingham-57763362, accessed August 8, 2021.

18. See Russell's principles of Beneficial AI, of which one is "purely altruistic machines" (2019, pp. 173–175).

19. Lee, 2018, p. 196.

20. Harari, 2018, p. 356. Irish author and journalist Mark O'Connell describes Kurzweil's singularity as "a kind of computational pantheism, a reverence for nature as an expression of a universal machine" (2017, loc. 1102).

21. "De-repress spiritual intelligence" is a phrase I first encountered in the writings of Ken Wilber (2011, p. 191).

22. Scharmer, 2016 (also for Buber); Rohr, 2012; Laloux, 2014; and Bourgeault, 2011.

23. Scharmer, 2016, loc. 3887.

24. Bienkowski, 2010, pp. 36–39.

25. Hillis in Brockman (ed.), 2019, pp. 173–177.

EPILOGUE

1. Gorman, 2021.

APPENDIX 2

1. "What Is Artificial Intelligence" (http://www-formal.stanford.edu/jmc/whatisai.pdf, updated November 12, 2007, accessed October 26, 2020). Two other definitions of AI are: "The ability of machines to perform tasks that normally require human intelligence" (www.afresearchlab.com, accessed August 6, 2021) and "AI is about building machines that do the right thing, that act in ways that can be expected to achieve their objectives" (Norvig and Russell, 1994).

2. Lee, 2018, p. 10.

3. Tegmark, 2018, p. 37.

4. This definition is close to Nick Bostrom's, "an intellect that is much smarter than the best human brains in practically every field, including scientific creativity, general wisdom and social skills" (https://www.nickbostrom.com/superintelligence.html, accessed October 25, 2020).

5. See Scharmer, 2016, loc. 312.

6. On philosophy and logic (including including a brief discussion of Parminedes, Heidegger, Plato), see Blackburn, 1996, p. 40. On Buddhist, Islamic, Neo-Daoist, and Zoroastrian treatments of being, see Carr and Mahalingam (eds.), 2000, pp. 576-577 and 770–771,

924–925 and 978–997, 577, and 11–12 respectively. See also Arthur Lovejoy's "Great Chain of Being," in which humanity is one link in the chain of being, along with plants, animals, and God (Lovejoy, 2009).

7. The Oxford English Dictionary provides the following four definitions of "being": existence; the nature or essence (of a person, etc.); a human being, and anything that exists or is imagined (Oxford English Reference Dictionary, 1995).

APPENDIX 3

1. This appendix is informed by the following sources, in addition to my own experience and reflection: A conversation with Joseph Chapa, June 9, 2021; "The Ethics of Artificial Intelligence," a McKinsey webcast with Michael Chui and Roger Burkhardt, February 26, 2019; and https://www.theguardian.com/technology/2021/apr/04/online-games-ai-emotion-recognition-emojify (April 4, 2021, accessed April 4, 2021).

2. Https://research.aimultiple.com/ai-ethics/, accessed July 8, 2021.

3. Http://www.europarl.europa.eu/doceo/document/A-8-2017-0005_EN.html, accessed September 20, 2020.

4. Kissinger, 2018.

5. Https://research.aimultiple.com/ai-ethics/, accessed July 8, 2021.

6. See https://www.technologyreview.com/2020/07/17/1005396/predictive-policing-algorithms-racist-dismantled-machine-learning-bias-criminal-justice/ (accessed July 14, 2021) and *Predictive Policing and Artificial Intelligence* (1st ed.), J. McDaniel and K. Pease (eds.), Routledge, 2021.

7. For a discussion of risks such as implementing algorithms that businesses could have been expected to know were discriminatory, see https://www.ibe.org.uk/uploads/assets/5f167681-e05f-4fae-ae1bef7699625a0d/ibebriefing58businessethicsandartificialintelligence.pdf, accessed August 23, 2021.

APPENDIX 4

1. See Wolfram in Brockman (ed.), 2019, p. 284.

2. Ibid., p. 282.

3. See discussion in chapter 4. In a different place, Wolfram may have included a form of purpose close to intentional purpose in his thinking. Brockman quoted Wolfram as follows: "I see technology as taking human goals and making them automatically executable by machines . . . the inventing of goals is not something that has a path to automation" (Brockman [ed.], 2019, p. 268) This is a recognition that human goals differ from technological goals.

4. Wolfram (in Brockman [ed.]) 2019, pp. 281–283.

5. Https://www.theatlantic.com/magazine/archive/2018/06/henry-kissinger-ai-could-mean-the-end-of-human-history/559124/, accessed July 14, 2021. At one point in this article, Kissinger discussed AlphaZero's victories at the game of Go. Kissinger observed that AI "establishes its own objectives," before going on to state that in respect of Go, "AI knows only one purpose: to win." Humans gave AlphaZero its purpose of winning; AlphaZero did not establish this objective for itself. Maybe Kissinger was distinguishing between purpose and objective, but he did not define any such difference in this article.

6. Https://mathworld.wolfram.com/GameofLife.html, accessed July 12, 2021.

7. Wac.36f4.edgecastcdn.net/0036F4/pub/www.wolframscience.com/nks/nks-ch11-sec2.pdf, accessed March 29, 2019.

8, Tallis, 2013, p. 759.

9. Some computer science engineers accept that human purpose and algorithmic purpose are not the same, but then claim that we should not put purpose of any kind into machines. However, they struggle to implement this claim. For example, Stuart Russell wrote that "we should avoid 'putting a purpose into the machine', as [mathematician] Norbert Wiener put it" (Russell, 2019, p. 203). Wiener seems to have coined his phrase with mechanistic purpose in mind (Pinker in Brockman [ed.], 2019, p. 103. Pinker notes that Norbert Wiener did not explain the personal aspect of purpose, but then gives Wiener too much credit for "explaining the hitherto mysterious world of purposes" in the 1950s. I am unsure how someone can explain such a world while omitting half of the territory). However, in a nuanced discussion, Russell then advocated putting some combination of intentional and mechanistic purpose into machines, as he conceives of machines satisfying human preferences according to his Principles of Beneficial AI. (See, for example, Russell, 2019, pp. 53 and 209–210 and Yudkowsky, 2018.) Russell's treatment of purpose recognizes that human preferences are uncertain and often driven by emotion (2019, pp. 231–235). For an interesting technical perspective on intentional purpose applied to eleven proposals for building safe advanced AI, see "An Overview of 11 Proposals for Building Safe Advanced AI," an article by Evan Hubinger Research Fellow, Machine Intelligence Research Institute, May 29, 2020 (cited as arXiv:2012.07532; https://arxiv.org/pdf/2012.07532.pdf, accessed August 3, 2021). Of the four criteria Hubinger uses to evaluate the eleven proposals (outer alignment, inner alignment, training competitiveness, and performance competitiveness) the last one is adjacent to the idea of intentional purpose. In order to work well with this uncertainty, leaders will benefit from recognizing that intentional purpose remains relevant. Threshold leaders create environments that invite each person's human purpose. They also pause before assuming that AIs will significantly guide organizational purpose, vision, or mission statements, since algorithms lack purpose of the kind that humans possess.

APPENDIX 5

1. Https://hbr.org/2005/04/seven-transformations-of-leadership, accessed July 30, 2021.

2. Garvey Berger, 2012, pp. 21–23.

3. Eigel and Kuhnert, 2005.

4. Ibid.

5. Josephs and Joiner, 2016, and http://integralleadershipreview.com/5535-fresh-perspective-leadership-agility-with-bill-joiner-and-steve-josephs/, accessed July 29, 2021.

6. Josephs and Joiner, 2016, p. 231.

7. Ibid., pp. v and 8–9.

8. Professor Rick Reis summarized the research as follows: "A four-year longitudinal study of twenty-two adults conducted by Kegan, Lahey, Souvaine, Popp, and Beukema using the Subject-Object Interview (Lahey, Souvaine, Kegan, Goodman, & Felix, 1988) revealed that 'at any given moment, around one-half to two-thirds of the adult population appears not to have fully reached the fourth order of consciousness" (Kegan, 1994, pp. 188, 191). Drawing on thirteen other studies conducted mainly by his doctoral students, Kegan (1994) reported that in the composite sample of 282 relatively advantaged adults, 59 percent had not reached

the fourth order. Findings from a longitudinal study of identity development of West Point cadets using Kegan's (1982, 1994) theory as a framework indicated that for most cadets, the challenge of college is moving from self-interest (order 2) to thinking in terms of being part of a community (order 3), a goal that must be accomplished before self-authorship can be considered (Lewis et al., 2005), https://tomprof.stanford.edu/posting/1110, accessed August 8, 2021.

9. Kegan, 1994.

BIBLIOGRAPHY

Aizenman, J., M. Lee, and D. Park. "The Relationship Between Structural Change and Inequality: A Conceptual Overview with Special Reference to Developing Asia." ADBI Working Paper, 396, Tokyo, Japan: Asian Development Bank Institute, 2012. http://www.adbi.org/workingpaper/2012/11/13/5332.structural.change.inequality.dev.asia/.

Anderson, Robert J., and William A. Adams. *Scaling Leadership: Building Organizational Capability and Capacity to Create Outcomes that Matter Most*. New Jersey, USA: Wiley, 2019. Kindle.

—*Mastering Leadership: An Integrated Framework for Breakthrough Performance and Extraordinary Business Results*. Hoboken, NJ, USA: Wiley, 2015. Kindle.

Babic, Boris, Daniel L. Chen, Theodoros Evgeniou, and Anne-Laure Fayard. "A Better Way to Onboard AI." *Harvard Business Review,* Volume 98, Issue 4, July–August 2020**,** pp. 56–65.

Baker, Greg. *The Energy Equation: Unlocking the Hidden Power of Energy in Business*. Hoboken, NJ, USA: Wiley, 2019. Kindle.

Basseches, Michael, and Michael F. Mascolo. *Psychotherapy as a Developmental Process*. Abingdon-on-Thames, UK: Routledge, 2009. Kindle.

Berinato, Scott. "That Discomfort You're Feeling Is Grief." *Harvard Business Review*, March 23, 2020. https://hbr.org/2020/03/that-discomfort-youre-feeling-is-grief, accessed March 20, 2022.

Berger, Peter L. *A Rumour of Angels*. Westminster, UK: Penguin, 1969. Print.

Bienkowski, Andrew. *One Life to Give*. New York, USA: The Experiment LLC, 2010. Print.

Blackburn, Simon. *Think*: *A Compelling Introduction to Philosophy*. Oxford, UK: Oxford University Press, 1999. Print.

Blakemore, Colin. *The Mind Machine*. London: BBC Books, 1990. Print.

Bostrom, Nick. "Ethical Issues in Advanced Artificial Intelligence." Previous published as: "Cognitive, Emotive and Ethical Aspects of Decision Making in Humans and in Artificial Intelligence." International Institute of Advanced Studies in Systems Research and Cybernetics, Vol. 2, ed. 1, 2003, pp. 12–17. https://www.nickbostrom.com/ethics/ai.html, accessed November 1, 2019.

Bourgeault, Cynthia. *The Wisdom Jesus: Transforming Heart and Mind—A New Perspective on Christ and His Message*. Boston, USA: Shambhala, 2009.

—*Superintelligence: Paths, Dangers, Strategies*. Oxford, UK: Oxford University Press, 2014. Kindle.

Brockman, John, editor. *Possible Minds: Twenty-Five Ways of Looking at AI*. New York, USA: Penguin Press, 2019. Print.

Brown, Brené. *The Gifts of Imperfection: Let Go of Who You Think You're Supposed to Be and Embrace Who You Are*. Hazelden, MN, USA: Hazelden Publishing, 2010. Print.

Buolamwini, Joy, and Timnit Gebru. "Gender Shades: Intersectional Accuracy Disparities in Commercial Gender Classification." *Proceedings of Machine Learning Research*, 81;1–15, 2018.

Campbell, Joseph. *The Hero with a Thousand Faces,* 3rd ed. Novato, CA, USA: New World Library, 2008.

Carr, Brian, and Indira Mahalingam, editors. *Companion Encyclopaedia of Asian Philosophy*. London, UK: Routledge, 2000. Print.

Chalmers, David. *The Conscious Mind: In Search of a Fundamental Theory,* 2nd edition. Oxford, UK: Oxford University Press, 1997.

Chapman, Gary. *The Five Love Languages: How to Express Heartfelt Commitment to Your Mate*. Chicago, USA: Northfield Publishing, 1992. Print.

Chatrath, Nick. "Fighting the Unbeliever: Anjem Choudary, Musharraf Hussain, and Pre-Modern Sources on Sūra 9.29, Abrogation and Jihad," *Islam and Christian-Muslim Relations*, 21: 2, 111–126.

Christianson, S. A., and E. Loftus. "Remembering Emotional Events: The Fate of Detailed Information." *Cognition and Emotion*, 5 (1991).

Conn, Charles, and Robert McLean. "Six Problem-Solving Mindsets for Very Uncertain Times." *McKinsey Quarterly*, September 2020.

Craig, Edward, editor. *The Shorter Routledge Encyclopedia of Philosophy*. New York, USA: Routledge, 2005. Print.

Csikszentmihalyi, Mihaly. *Flow*. London, UK: Random House, 1992. Print.

Dennett, Daniel C. "What Can We Do?" In Brockman, editor. *Possible Minds: Twenty-Five Ways of Looking at AI*. New York, USA: Penguin Press, 2019, pp. 41–53.

Deutsch, David. "Beyond Reward and Punishment." In Brockman, editor. *Possible Minds: Twenty-Five Ways of Looking at AI*. New York, USA: Penguin Press, 2019, pp. 115–124.

Dihal, Kanta, Sarah Dillon, and Beth Singler. "From Homer to Hal." *Research Horizons* (University of Cambridge), Issue 35, pp. 28–29, February 2018. Print.

Dirckx, Sharon. *Am I Just My Brain?* Epsom, UK: The Good Book Company, 2019. Print.

Draper, Brian. *Spiritual Intelligence: A New Way of Being*. Oxford, UK: Lion Books, 2011. Print.

Eigel, Keith M., and Karl W. Kuhnert. "Authentic Development: Leadership Development Level and Executive Effectiveness." *Authentic Leadership Theory and Practice: Origins, Effects and Development Monographs in Leadership and Management*, Volume 3 (2005), pp. 357–385.

Ekman, Paul. "An Argument for Basic Emotions." *Cognition and Emotion*, 6 (1992), pp. 169–200.

Epley, Nicholas, and Amit Kumar. "How to Design an Ethical Organization." *Harvard Business Review*, Volume 97, Issue 3, May–June 2019, pp. 144ff.

Evans, Dylan. *Emotion: A Very Short Introduction*. Oxford, UK: Oxford University Press, 2001. Print.

Feldman Barrett, Lisa. *How Emotions Are Made: The Secret Life of the Brain*. New York, USA: Macmillan, 2017. Kindle.

Fry, Hannah. *Hello World: How to Be Human in the Age of the Machine*. London, UK: Penguin, 2018. Kindle.

Gardner, Howard. *Multiple Intelligences: New Horizons in Theory and Practice*. New York, USA: Basic Books, 2006. Kindle.

Garvey Berger, Jennifer. *Changing on the Job: Developing Leaders for a Complex World*. Redwood City, CA, USA: Stanford University Press, 2012. Kindle.

Gino, Franscesca. "The Business Case for Curiosity." *Harvard Business Review*, Volume 96, Issue 5, September–October 2018, pp. 48–57.

Gladwell, Malcolm. *Blink: The Power of Thinking Without Thinking*. London, UK: Penguin Books, 2005. Print.

Goetzel, Ben. "Superintelligence: Fears, Promises, and Potentials." *Journal of Evolution and Technology*, Volume 25, Issue 2, 2015, pp. 55–87. Accessed on January 31, 2019, https://jetpress.org /v25.2/goertzel.htm; also available at http://www.kurzweilai.net/superintelligence-fears-promises -and-potentials.

Goleman, Daniel. *Emotional Intelligence: Why It Can Matter More Than IQ*. London, UK: Bloomsbury, 1995. Print.

Gopnik, Alison. "AIs Versus Four-Year-Olds." In Brockman, editor. *Possible Minds: Twenty-Five Ways of Looking at AI*. New York, USA: Penguin Press, 2019, pp. 221–230.

Gorman, Amanda. *The Hill We Climb: An Inaugural Poem for the Country*. New York, USA: Viking Books, 2021. Print.

Harari, Yuval Noah. *21 Lessons for the 21st Century*. London, UK: Vintage Books. 2018. Kindle.

—*Homo Deus: A Brief History of Tomorrow*. New York, USA: Vintage Digital. 2016. Kindle.

Harris, Sam, and Eliezer Yudkowsky. "AI: Racing Toward the Brink." *Making Sense* podcast (Sam Harris) Podcast, Episode #116, February 6, 2018.

Hawking, Stephen. "Will Artificial Intelligence Outsmart Us?" In *Brief Answers to the Big Questions*. London, UK: John Murray, 2018, pp. 181–196. Print.

Henderson-Espinoza, Robyn. *Body Becoming: A Path to Our Liberation*. Minneapolis, MN, USA: Broadleaf Books, 2022. Kindle.

Hillis, W. Daniel. "The First Machine Intelligences." In Brockman, editor. *Possible Minds: Twenty-Five Ways of Looking at AI*. New York, USA: Penguin Press, 2019, pp. 171–80.

Jacobs, Alan. *How to Think: A Guide for the Perplexed*. London, UK: Profile Books, 2017. Kindle.

Josephs, Stephen A., and William B. Joiner. *Leadership Agility: Five Levels of Mastery for Anticipating and Initiating Change*. San Francisco, USA: Jossey-Bass, 2016. Kindle.

Kahneman, Daniel. *Thinking, Fast and Slow*. London, UK: Penguin Books, 2011. Kindle.

Kasparov, Garry. *Deep Thinking: Where Machine Intelligence Ends and Human Creativity Begins*. London, UK: John Murray, 2017. Kindle.

Kegan, Robert. *In Over Our Heads: The Mental Demands of Modern Life*. Cambridge, MA, USA: Harvard University Press, 1994. Print.

—*The Evolving Self: Problems and Process in Human Development*. Cambridge, MA, USA: Harvard University Press, 1982. Print.

Kegan, Robert, and Lisa Lahey. *An Everyone Culture: Becoming a Deliberately Developmental Organization*. Boston, USA: Harvard Business School Press, 2016. Kindle.

—*Immunity to Change: How to Overcome it and Unlock the Potential in Yourself and Your Organization*. Boston, USA: Harvard Business School Press, 2009. Print.

—*How the Way We Talk Can Change the Way We Work: Seven Languages for Transformation*. San Francisco, USA: Jossey-Bass, 2001. Print.

Keller, Scott, and Bill Schaninger. *Beyond Performance 2.0: A Proven Approach to Leading Large-Scale Change*. Hoboken, NJ, USA: Wiley, 2019. Print.

Kelly, Kevin. *The Inevitable: Understanding the 12 Technological Forces That Will Shape Our Future*. New York, USA: Penguin, 2017. Kindle.

Kessler, David. *Finding Meaning: The Sixth Stage of Grief.* London, UK: Ebury Digital, 2019. Kindle.

Kissinger, Henry, "How the Enlightenment Ends," *The Atlantic*, June 2018 Issue, https://www.theatlantic.com/magazine/archive/2018/06/henry-kissinger-ai-could-mean-the-end-of-human-history/559124/.

Kline, Nancy. *The Promise That Changes Everything: I Won't Interrupt You.* London, UK: Penguin Random House, 2020. Print.

—*More Time to Think.* London, UK: Octopus Publishing Group Ltd., 2015. Print.

—*Time to Think.* London, UK: Octopus Publishing Group Ltd., 2011. Kindle. (Print edition: London: Cassell, 1999.)

Kurzweil, Ray. *How to Create a Mind.* London, UK: Duckworth, 2013. Kindle.

Laloux, Frederic. *Reinventing Organizations: A Guide to Creating Organizations Inspired by the Next Stage of Human Consciousness.* Brussels, BE: Nelson Parker, 2014. Kindle.

Lee, Kai-Fu. *AI Superpowers: China, Silicon Valley, and the New World Order.* Boston, USA: Houghton Mifflin Harcourt, 2018. Kindle.

Legg, Shane. *Machine Super Intelligence*, a doctoral dissertation submitted to the Faculty of Informatics of the University of Lugano, 2008. Accessed under the Creative Commons license: http://www.vetta.org/documents/Machine_Super_Intelligence.pdf.

Lencioni, Patrick. *The Advantage.* San Francisco, USA: Jossey-Bass, 2012. Kindle.

Lennox, John. *2084.* Grand Rapids, MI, USA: Zondervan Reflective, 2020. Kindle.

—*Can Science Explain Everything?* Epsom, UK: The Good Book Company, 2019. Kindle.

Lloyd, Seth. "Wrong, but More Relevant Than Ever." In Brockman, editor. *Possible Minds: Twenty-Five Ways of Looking at AI.* New York, USA: Penguin Press, 2019, pp. 3–12.

Lovejoy, Arthur. *The Great Chain of Being: A Study of the History of an Idea.* Cambridge, MA, USA: Harvard University Press, 2009. Print.

MacFarquhar, Larissa. "The Mind-Expanding Ideas of Andy Clark." *New Yorker Annals of Thought*, March 26, 2018.

Malloch, Theodore Roosevelt. *Practical Management Wisdom: Business Across Spiritual Traditions.* Sheffield, UK: Greenleaf Publishing, 2015. Print.

Malnight, Thomas W., Ivy Buche, and Charles Dhanaraj. "Put Purpose at the Core of Your Strategy." *Harvard Business Review*, Volume 97, Issue 5, September–October 2019, pp. 70–79.

Martin, Roger. *The Design of Business: Why Design Thinking Is the Next Competitive Advantage.* Boston, USA: Harvard Business School Press, 2009. Kindle.

McBride, Hillary L. *The Wisdom of Your Body: Finding Healing, Wholeness, and Connection Through Embodied Living.* Grand Rapids, MI, USA: Brazos Press, 2021. Kindle.

McGilchrist, Ian. *The Matter with Things: Our Brains, Our Delusions and the Unmaking of the World.* London, UK: Perspectiva Press, 2021. Kindle.

McGrath, Rita, and Ryan McManus. "Discovery-Driven Digital Transformation." *Harvard Business Review*, Volume 98, Issue 3, May–June 2020, pp. 124–133.

McLaren, Brian D., and Gareth Higgins. *The Seventh Story: Us, Them, & the End of Violence.* 2019. Kindle.

Minto, Barbara. *The Pyramid Principle.* Edinburgh, UK: Pearson Education Ltd, 1987. Print.

Monti, M. M., A. Vanhaudenhuyse, M. R. Coleman, M. Boly, J. D. Pickard, J. F. L. Tshibanda, A. M. Owen, and S. Laureys. "Willful Modulation of Brain Activity and Communication in Disorders of Consciousness." *New England Journal of Medicine*, Volume 362, 2010, pp. 579–589.

Moser J. S., H. S. Schroder, C. Heeter, T. P. Moran, and Y. H. Lee. "Mind Your Errors: Evidence for a Neural Mechanism Linking Growth Mind-Set to Adaptive Posterior Adjustments." *Psychol. Sci.*, Volume 22, 2011, pp. 1484–1489.

Norvig, Peter, and Stuart Russell. *Artificial Intelligence: A Modern Approach*. Hoboken, NJ, USA: Prentice Hall, 1994. Print.

Oakley, Mark. *The Splash of Words: Believing in Poetry*. London, UK: Canterbury Press, 2017. Print.

Obrist, Hans Ulrich. "Making the Invisible Visible: Art Meets AI." In Brockman, editor. *Possible Minds: Twenty-Five Ways of Looking at AI*. New York, USA: Penguin Press, 2019, pp. 208–218.

O'Connell, Mark. *To Be a Machine: Adventures Among Cyborgs, Utopians, Hackers, and the Futurists Solving the Modest Problem of Death*. London, UK: Granta, 2017. Kindle.

Ogden, Pat, and Janina Fisher. *Sensorimotor Psychotherapy: Interventions for Trauma and Attachment*. New York, USA: Norton, 2015. Kindle.

Okasha, Samir. *Philosophy of Science: A Very Short Introduction*. Oxford, UK: Oxford University Press, 2016. Print.

Orford, Pete Robert. "Dickens and Science Fiction: A Study of Artificial Intelligence in *Great Expectations*." *Interdisciplinary Studies in the Long Nineteenth Century*, Volume 10, 2010.

Owen, Adrian. *Into the Grey Zone: A Neuroscientist Explores the Border Between Life and Death*. London, UK: Guardian Faber Publishing. 2017. Kindle.

Owen, A. M., M. R. Coleman, M. H. Davis, M. Boly, S. Laureys, J. D. Pickard. "Detecting Awareness in the Vegetative State." *Science*, Volume 313, 2006, p. 1402.

Oxford Reference. *The Oxford Dictionary of Philosophy*. Oxford, UK: Oxford University Press, 1996. Print.

Palmer, Parker J. *Let Your Life Speak: Listening for the Voice of Vocation*. San Francisco, USA: Jossey-Bass, 2000. Print.

Pearl, Judea. "The Limitations of Opaque Learning Machines." In Brockman, editor. *Possible Minds: Twenty-Five Ways of Looking at AI*. New York, USA: Penguin Press, 2019, pp. 15–19.

Pentland, Alex. "'Sandy,' The Human Strategy." In Brockman, editor. *Possible Minds: Twenty-Five Ways of Looking at AI*. New York, USA: Penguin Press, 2019, pp. 194–205.

Pink, Daniel. *A Whole New Mind*. London, UK: Marshall Cavendish, 2008. Print.

Pinker, Steven. "Tech Prophecy and the Underappreciated Causal Power of Ideas." In Brockman, editor. *Possible Minds: Twenty-Five Ways of Looking at AI*. New York, USA: Penguin Press, 2019, pp. 101–112.

Radecki, Dan, and Leonie Hull (with Jennifer McCusker and Christopher Ancona). *Psychological Safety: The Key to Happy, High-Performing People and Teams*. Orange County, CA, USA: Academy of Brain-Based Leadership, 2018. Kindle.

Ramakrishnan, Venki. "Will Computers Become Our Overlords." In Brockman, editor. *Possible Minds: Twenty-Five Ways of Looking at AI*. New York, USA: Penguin Press, 2019, pp. 181–191.

Riegart, Ray. and Thomas Moore, editors. *The Lost Sutras of Jesus: Unlocking the Ancient Wisdom of the Xian Monks*, 2003. Print.

Roberts, Robert C., and W. Jay Wood. *Intellectual Virtues: An Essay in Regulative Epistemology*. Oxford, UK: Clarendon Press, 2009. Print.

Rock, David. *Your Brain at Work: Strategies for Overcoming Distraction, Regaining Focus, and Working Smarter All Day Long*. New York, USA: HarperCollins, 2009. Kindle.

Rohr, Richard. *Immortal Diamond*. London, UK: SPCK, 2012. Print.

—*Falling Upward: A Spirituality for the Two Halves of Life*. San Francisco, USA: Jossey-Bass, 2012. Kindle.

Russell, Stuart. *Human Compatible: Artificial Intelligence and the Problem of Control*. New York, USA: Viking, 2019. Print.

Ryff, Carol, and Corey Lee M. Keyes. "The Structure of Psychological Well-Being Revisited." *Journal of Personality and Social Psychology*, 1995, Volume 69, No. 4, pp. 717–719.

Sanders, Scott Russell. "Under the Influence." In *The Norton Reader* (shorter 13th edition), 2012, pp. 60–70.

Scharmer, Otto. *Theory U: Leading from the Future As It Emerges*. Oakland, CA, USA: Berrett-Koehler Publishers, 2016. Kindle.

Schutz, Will. *Profound Simplicity*. New York, USA: Bantam Books, 1979. Print.

Schwartz, Tony, Jean Gomes, and Catherine McCarthy. *The Way We're Working Isn't Working: The Four Forgotten Needs That Energise Great Performance*. London, UK: Simon & Schuster, 2010. Print.

Sternberg, Robert J. *Beyond IQ: A Triarchic Theory of Human Intelligence*. New York, USA: Cambridge University Press, 1985. Print.

Tallis, Raymond. *Aping Mankind: Neuromania, Darwinitis and the Misrepresentation of Humanity*. Durham, NC, USA: Acumen, 2012. Print.

—*Why the Mind Is Not a Computer: A Pocket Lexicon of Neuromythology*. Exeter, UK: Andrews UK Limited. 2013. Digital edition.

Taylor, Sonya Renee. *The Body Is Not an Apology*, 2nd edition. Oakland, CA, USA: Berrett-Koehler, 2021. Kindle.

Tegmark, Max. "Let's Aspire to More Than Making Ourselves Obsolete." In Brockman, editor. *Possible Minds: Twenty-Five Ways of Looking at AI*. New York, USA: Penguin Press, 2019, pp. 78–87.

—*Life 3.0: Being Human in the Age of Artificial Intelligence*. London, UK: Penguin Books, 2018. Kindle.

—*Our Mathematical Universe*. London, UK: Penguin Books, 2015.

Turner, Victor. *The Ritual Process: Structure and Anti-Structure*. New York, USA: Cornell University Press, 1969. Print.

Van der Kolk, Bessel. *The Body Keeps the Score: Brain, Mind, and Body in the Transformation of Trauma*. London, UK: Penguin Books, 2014. Kindle.

Waldock, Trevor. *To Plant a Walnut Tree: How to Create a Fruitful Legacy by Using Your Experience*. Boston, USA: Nicholas Brealey Publishing, 2011. Print.

Walker, Matthew. *Why We Sleep: The New Science of Sleep and Dreams*. London, UK: Penguin Books, 2018. Print.

Watts, Isaac. *The Improvement of the Mind*. Elibron Classics, Adamant Media Corporation, 2005. Facsimile of the 1837 edition. Print.

Whyte, David. *The Heart Aroused: Poetry and the Preservation of the Soul in Corporate America*. New York, USA: Doubleday, 1994. Print.

Wilber, Ken. *Integral Spirituality*. Boston, USA: Integral Books, 2011. Kindle.

Wilczek, Frank. "The Unity of Intelligence." In Brockman, editor. *Possible Minds: Twenty-Five Ways of Looking at AI.* New York, USA: Penguin Press, 2019, pp. 66–75.

Williams, Mark, and Danny Penman. *Mindfulness: A Practical Guide to Finding Peace in a Frantic World.* London, UK: Piatkus, 2011. Kindle.

Wilson, H. James, and Paul R. Daugherty. "Collaborative Intelligence: Humans and AI Are Joining Forces." *Harvard Business Review*, Volume 96, Issue 4, July–August 2018, pp. 114–123.

Winnicott, D. Chapter 12. "Ego Distortion in Terms of True and False Self." In *The Maturational Process and the Facilitating Environment: Studies in the Theory of Emotional Development.* New York, USA: International Universities Press, Inc., pp. 140–157.

Wiseman, Liz. *Multipliers: How the Best Leaders Make Everyone Smarter.* New York, USA: Harper Business, 2017 (revised and updated edition). Kindle.

Wolfram, Stephen. "Artificial Intelligence and the Future of Civilization." In Brockman, editor. *Possible Minds: Twenty-Five Ways of Looking at AI.* New York, USA: Penguin Press, 2019, pp. 268–284.

Wooldridge, Mike. "The Two Routes to Artificial Intelligence." *Discover #OxfordAI,* 2nd ed., 2019, https://www.research.ox.ac.uk/Article/2019-02-28-discover-oxfordai-brochure, accessed November 1, 2019.

Yudkowsky, Eliezer. "AI: Racing Toward the Brink." *Making Sense* podcast (Sam Harris), Ep. 116, February 6, 2018.

Zohar, Danah. *Spiritual Intelligence: The Ultimate Intelligence.* London, UK: Bloomsbury Books. Print.

Zuboff, Sandra. *The Age of Surveillance Capitalism.* London, UK: Profile Books, 2019. Print.

INDEX

ABOUT THE AUTHOR

Nick Chatrath has dedicated his professional life to helping leaders flourish. A former McKinsey consultant, he has spent more than twenty thousand hours coaching executives, educators, politicians, diplomats, military leaders, home-makers, students, and others. Nick has seen leaders from all sectors and from all walks of life as they have struggled and claimed great victories.

Nick has held several technology leadership roles and cherishes diverse approaches to it. He cofounded two technology startups, one of which was AI-integrated. Since arriving at university for the first time, Nick has studied mathematics, management studies, French, theology (including modules on philosophy), history, Arabic, and Islamic law at the bachelor's, master's, or doctoral level, including at Oxford and Cambridge. Nick is passionate about integrating all these disciplines in the context of technology and leadership. Nick coleads Artesian Transformational Leadership, a global professional services firm focusing on leadership development.

Half Greek, half Indian, and raised in the UK, Nick has traveled and worked extensively around the world. He lives in Oxford with his wife and three daughters. In his spare time, Nick competes in sprint triathlons.